Victor Vecki

The Pathology and Treatment of Sexual Impotence

Second Edition

Victor Vecki

The Pathology and Treatment of Sexual Impotence
Second Edition

ISBN/EAN: 9783337280161

Printed in Europe, USA, Canada, Australia, Japan

Cover: Foto ©berggeist007 / pixelio.de

More available books at **www.hansebooks.com**

THE PATHOLOGY AND TREATMENT

OF

SEXUAL IMPOTENCE

BY

VICTOR G. VECKI, M.D.

FROM THE AUTHOR'S SECOND GERMAN
EDITION, REVISED AND REWRITTEN

PHILADELPHIA
W. B. SAUNDERS
925 Walnut Street
1899

PREFACE.

WHEN the first German edition of this work was published, in 1889, there was some commotion in the ranks of old and young medical fogies, who were indignant that any one dared to resist their intellectual tendencies, refused to worship their superannuated gods.

The second German edition found the ranks of the same kind of professional formula-riders and bigots solid, though somewhat thinned.

I have taken the liberty of preserving the independence of my altruistic opinions, and shall continue to fight against false and hypocritical quasi-scientific pretensions. The circumstance that my work has been given earnest consideration by authorities like Casper, Eulenburg, Fürbringer, Krafft-Ebing, and others makes it easy to bear all the acrimonious aggressions dictated by the bilious nature of some of the "Dii minorum gentium."

I wish to thank Professor A. A. D'Ancona, who kindly revised the manuscript and helped me in many other ways.

THE AUTHOR.

22 GEARY STREET,
    SAN FRANCISCO, CAL.

PREFACE
TO THE SECOND GERMAN EDITION.

In the lapse of seven years passed since the publication of the first edition of this work, we can record but very little progress in the theoretic as well as the practical development of our subject. We know to-day just as much—or, better, just as little—about the physiology of the sexual act as we did seven years ago.

Quite new, indeed, is an abundance of newly-forged names for old pathologic conditions. Some authors try to perpetuate themselves in this way. We can only hope that most of these new names will be short-lived.

The therapeutics of sexual impotence has received some valuable additions, and we have in the method of suspensions a frequently efficacious, and in hypnotic suggestion an occasionally efficacious, remedy.

It affords me special satisfaction that my monograph has not proved to be an ephemera, in spite of the many adversaries the liberal interpretation of some pertinent questions has encountered.

THE AUTHOR.

SAN FRANCISCO, CAL., November, 1896.

CONTENTS.

CHAPTER	PAGE
I.—Introduction	17
II.—Anatomy	30
III.—Physiology of the Sexual Act	45
Sexual Maturity	45
Sexual Orgasm	47
Seat of the Sexual Instinct	50
Erection	51
Ejaculation	59
The Semen	61
IV.—Etiology of Impotence	79
V.—Forms of Impotence	87
Congenital Malformations and Defects of the Sexual Organs	87
Absence of Penis	87
Diminutive Size of Penis	88
Excessive Size of Penis	88
Absence of Prepuce	89
Superfluity of Prepuce	90
Hypospadiasis	91
Epispadiasis	91
Monorchis	92
Cryptorchis	92
Hermaphrodites	93

CHAPTER	PAGE
Acquired Defects in the Organs of Generation	93
Absence of Penis and Testicles	93
Absence of Testicles	94
Absence of One Testicle	95
Hydrocele and Inguinal Hernia	95
Changes in the Corpora Cavernosa	96
Consecutive Impotence	97
Acute Diseases	97
Phthisis	98
Chronic Diseases	99
Diabetes	99
Obesity	99
Anemia	100
Diseases of Brain and Spinal Cord	102
Neurasthenia	103
Diseases of the Genitalia	108
Alcohol	113
Coffee, etc.	116
Tobacco	116
Digitalis	117
Morphin, etc.	118
Arsenic	119
Lead	120
Iodin	120
Mercury	120
Salicylic Acid, Antipyrin, etc.	120
Inherited Predisposition to Impotence	122
Sexual Weakness	123
Incontinence of Urine	124
Frigidity	126
Nervousness	127
Perverse Sexual Feeling	128

CONTENTS.

CHAPTER	PAGE
Neurasthenic Impotence	136
Excess in Venery	137
After-Effects of Copulation	148
Sexual Excesses, Consequences of	152
Frigidity	153
Satiety of Ordinary Sexual Gratification	154
Paralytic Impotence	156
Onanism	160
Causes of Onanism	164
Consequences of Onanism	171
Nervous Diseases	172
Pollutions and Spermatorrhea	174
Endoscopic Examination, Results of	184
Abstinence	188
Irritable Weakness	192
Psychical Impotence	195
Temporary Impotence	198
Relative Impotence	199
Professional Impotence	202
Senile Impotence	205
VI.—DIAGNOSIS	209
VII.—PROGNOSIS	214
VIII.—PROPHYLAXIS	216
IX.—TREATMENT	226
Psychical Treatment	228
Removal of Morbid Influences	230
Treatment of Spermatorrhea	230

CHAPTER	PAGE
Hygiene of Living	236
Nourishment	237
Sleep	239
Medicaments	239
Cantharides, etc.	239
Phosphorus	240
Nux Vomica	241
Secale Cornutum	242
Quinin	242
Iron	242
Opiates	242
Valerian	243
Stimulants	243
Cocain	243
Scincus Marinus	243
Damiana	244
Hydrotherapeutics	244
Ablutions	246
Friction and Similar Proceedings	246
Sponge-Baths	247
Douches	247
Sitz-Baths	247
Half-Baths	248
Vapor-Baths	248
River- and Sea-Baths	248
Balneological Treatment	249
Psychrophor	249
Injections	250
Carbon Douche	250
Electro-Therapeutics	251
Galvanic Current	252

CHAPTER	PAGE

Electro-Therapeutics.—Continued.

    Faradic Current 253
    Static Electricity 254
    Hydro-Electric Bath 254

Local Treatment 254
    Cauterization 254
    Astringents 258
    Bougies and Sounds 260
    External Applications 261
    Sinapisms 261
    Acupuncture and Electropuncture .   . . 261
    Surgical Operations 261

Massage and Gymnastics 262

Travelling   . . 262

Flagellation 263

Apparatus and Instruments 264

Regulation of the Sexual Life 270

Matrimony 271

Inhalation of Oxygen 273

Suspensions 274

Hypnotic Suggestion 277

X.—SPECIAL THERAPEUTICS 280

SEXUAL IMPOTENCE.

CHAPTER I.

INTRODUCTION

To write on the much-scouted subject of sexual impotence is a venturesome undertaking under all circumstances; but to write without the customary affectation and hypocritical rolling of the eyes, to speak the bare truth, surely requires even greater courage. Many an eminent medical man may have felt a secret desire to take the risk, but refrained from carrying out the resolution through fear of endangering his professional reputation. Some one not counted among the magnates of the medical realm may feel licensed to plunge to the very bottom, "to see what the gods have covered with darkness and horror," and may dare to relate to his colleagues what he sees and hears, without alteration or retouching.

"No physical or moral suffering, no wound, however putrid it may be, should frighten him who devotes his life to the science of man; and the sacred ministry which obliges the physician to see everything, to know everything, gives him also the right to relate everything."[1]

---

[1] Tardieu, Étude médico-legale sur les attentats aux mœurs. Paris, 1878, p. 2.

It must be admitted that the subject has never received the attention its preëminent importance deserves. The world over it seems to be considered the proper thing to treat the affair with supercilious nonchalance. Few medical men in Germany can boast of ever having had an opportunity to hear a clinical lecture on impotency; and the complaints and criticisms of authors prove that elsewhere the subject receives no greater favor. "The subject is thus not yet emancipated from the tenacious grasp of the most rampant charlatanism."[1]

"Let us be frank from the first steps of our researches, because hypocrisy is the worm which in modern society attacks and corrodes the highest and most powerful plant of this life's garden."[2]

This somewhat serious neglect is no doubt to be attributed to the circumstances that those suffering from impotence can hardly be subjects for treatment in hospitals, and that the observation of the details and symptoms of the disease is attended with unusual difficulties; nay, is hardly possible at all.

Recent indications seem, however, to point to a better future. Men of prominence in the learned world, with Eulenburg, Krafft-Ebing, Fürbringer, Edw. Martin, and some others at the head, do not think it beneath their dignity to busy themselves with the solution of the perplexing problems of the sexual life, and it is to be hoped that before long the conventional medical lies with which every book, every pamphlet on the subject is swarming, will disappear, and Mantegazza's satire[3] become obsolete.

---

[1] Campbell Black, On the Functional Diseases of the Urinary and Reproductive Organs. London, 1875, p. 6.

[2] Mantegazza, Fisiologia dell'amore. Milano, 1882, p. 75.

[3] Op. cit., p. 298.

"Difficult problems cannot be solved if we run away from them or if we avoid them; and still many a physician, many a philosopher, tries to solve the most burning questions of modern society in the manner of the baby who believes that he can escape the threatening dog by closing his eyes."

No one denies that the sexual function is of very great consequence to the individual as well as to society in general, although one does not care to make this a subject of conversation.

"At any rate, the sexual function forms the most powerful factor in individual and in social life. It is a mighty impulse for bringing into action our most effective energies, for acquiring property, for the foundation of a home, for rousing altruistic feelings for a person of the other sex first, and, later, for one's children, and, in a wider sense, for the whole human family."[1] The civil code of Austria correctly estimates the importance of sexual virility, Article 60 declaring that "the continued inability to fulfil the conjugal duty is a bar to marriage." The criminal code (Section 156) declares, "But if the crime has caused the loss of the procreative power of the injured man, then the punishment of imprisonment with hard labor is to be meted out for from five to ten years."

Without virility there can be *no procreation*. That the semen of impotent men frequently contains spermatozoa need not be taken into account in considering the propagation of the race, and, generally speaking, there are certainly but very few men who owe their lives to impotent fathers.

---

[1] Krafft-Ebing, Psychopathia sexualis. Stuttgart, 1886, p. 2.

He who has become prematurely impotent is one of the most unfortunate creatures, his misfortune being the greater as he has to be ashamed of it, must conceal it, is pitied by no one, but scorned, and, alas, in exceedingly few cases can he hope for recovery. I venture to assert that in many cases it is a better deed to restore to an impotent man the power, so precious to every individual, than to preserve a dangerously sick person from death, for in many cases death is preferable to impotence.

The energy of man, his courage, his enjoyment of work and life, all, with hardly any exception, depend on his sexual power. Here I leave out of sight the sundry vows of chastity made by persons who expect to be rewarded in the hereafter for their voluntary martyrdom here below. These people hardly seem to be enjoying their lives, and they call this world a vale of tears. Now, this world is not exactly a vale of tears or of grief; but let a man who has to labor and produce from early morning till evening, who must day after day begin ever anew the struggle for existence, lose that little bit of love and his pathway will lead through a vale of tears indeed. The difference between the view of things in general formed by old people and that formed by the young has its only explanation in the virility extinguished in the former and vigorous in the latter. Prematurely impotent people very often appear aged physically, and always mentally. Moreover, we must not forget that the sexual nervous system is closely related to all the rest of the nervous mechanism, including those parts essential to its physiological operations. "The feeling of sexual impotence is the most humiliating which can ever afflict a man; because it degrades him in his own eyes, and does not leave a single possible illusion, not a solitary moment of mercy."[1] Eunuchs and the sexually impotent differ from their fellow-men who are in full possession of virile power in appearance and in conduct. Even though executed with talent and spirit, the work of an impotent man bears the stamp of impotence.

In the year 1878 I happened to be in the Paris Salon with one of the most famous French painters. While contemplating some paintings, the great master, to whose words a crowd of artists were eagerly listening, said, "That painter must be impotent." To a question of mine the master replied that he was able to tell by a picture, not only whether it was painted by a young man or an old one, but also the condition of the artist's sexual power. To-day I give full credit to this assertion, as I am convinced that with some experience one can distinguish an impotent man from a virile one merely by his looks, demeanor, ideas, words, and works.

The exterior of an incurably impotent man, whether his impotence be real or imaginary, does not always suggest physical weakness; on the contrary, many present a very healthy appearance, and are stout. The keen-eyed public have baptized this corpulency "capon-obesity." In spite of the apparent healthiness, the impotent man is generally melancholy, discontented, and peevish. The prematurely impotent are, without exception, ill-humored, cheered up with difficulty, and even then for but a short time. Most of them are grudging, cowardly, envious, and wicked. They are all very jealous, as may easily be understood. The younger they are, the handsomer their bodies, the higher their social rank, the more pronounced is their bad character.

---

[1] Lallemand, Pertes séminales, tome ii., 1re partie, p. 132.

The character of a man must, as a matter of course, be considerably affected by the consciousness of impotence. No one is more severe than the impotent in passing judgment on his neighbor. No one so ruthlessly or mercilessly condemns a misdeed, caused by passion, against the very wise prescripts of Ethics. Since he cannot join the virile in their enjoyments of life, he makes a merit of his incapacity.

The striving of a man to found for himself a home, a family, is a stimulus to work and to the accomplishment of great deeds in his sphere of life. Such a stimulus does not stir the impotent. Although they do not care for life, yet they are cowards. It is very seldom that an impotent man turns dare-devil and shows a contempt for death, due to despair.

The impotent are incapable of love; for, as Krafft-Ebing says, "With all the morality which love needs to rise to its true and pure character, its most vigorous root is nevertheless sexual passion. Platonic love is a non-entity, a self-deception, a wrong designation for cognate feelings."[1] Similarly, ambition is closely dependent upon the sexual power, as it seldom makes its appearance before puberty.

Finally, it must not be forgotten how wretched a part is played in matrimony and in every other relation to a woman by the man who is completely or even partially impotent. He must renounce the affection and regard of a woman. Galopin[2] is quite right in saying, "Without this good friend (the woman) the dawn and evening of life would be helpless, and its mid-day without pleas-

---

[1] Krafft-Ebing, Psychopathia sexualis. Stuttgart, 1886, p. 9.
[2] Le parfum de la femme. Paris, 1886, p. 101.

ure." With all the capacity for self-sacrifice which is inborn with the entire sex, women will nevertheless seldom soar to so high a degree of self-abnegation as to love an impotent man. It is the aim of every husband to hold a dominant position in his family; the more so as the weight of his voice is less elsewhere. My object is not to examine whether such a position is in the interest of the man himself and of his family, but I wish merely to call attention to the fact that the influence of an impotent man must be very insignificant with a woman living with him, whether she be wife or mistress.

I further wish to point out that many fallen women would have continued good and faithful had their husbands not been more or less impotent. The greater share of sexual appetite roused inconsiderately or ignorantly remains unsatisfied, and much matrimonial happiness is ruined by the husband's impotence. "Tutavia la compagnia fra moglie e marito si conferma grandemente per questo atto, e non pùo far miglior cosa il marito per tenersi affettionata e pacificata la moglie, che questa e spesso. Perchè a questa foggia, tutta la casa sta in pace, e tranquilla, e tutte le cose vanno bene." (Levinio Lennio.)[1]

Now and then impotence leads to suicide. Marc[2] tells, for instance, of a young man who, before committing suicide, had written down the words, "I am impotent, consequently I am good for nothing in this life."

If impotence declares itself slowly and gradually, refuge is not so often taken in this safe though extreme remedy for all diseases; but is applied more frequently

---

[1] Mantegazza, Igiene dell'amore. Milano, 1881, p. 95.
[2] Mantegazza, op. cit., p. 143.

when the calamity shows itself at once and without a
state of transition, so that the patient understands plainly
that there is no help for his ailment, and he has therefore not the time to accustom himself to his misfortune.
Richerand[1] made the observation that after penis ampu-
tation the patient becomes melancholy, and is consequently more subject to malignant wound fever, which
often causes death, whilst other mutilations by surgical
operations are borne with fortitude. Lallemand[2] tells
of a man forty-five years old who, after penis amputa-
tion, when on the point of leaving the hospital, received
a visit from his wife, after which he grew gloomy, mournful, taciturn, and died suddenly. The most careful autopsy failed to reveal any cause of death; Lallemand
ascribes the fatal result to despair. In this case the
patient had so much the more reason for despair, as the
sexual appetite had not vanished with the loss of the
penis. The Russian Skopzi after having been maimed
suffer a complete change of character: they grow ego-
istic, malicious, hypocritical, and covetous.[3] It is to be
observed, however, that surgical operations performed
on the genitals can cause genital reflex neuroses in the
form of melancholia.[4]

The fate of being impotent is borne with more stoicism
when along with the loss of virility goes that of every desire for intercourse with the other sex. Here again we
find in Lallemand[5] a typical example: A man about

---

[1] Roubaud, Traité de l'impuissance. Paris, 1876, p. 66.
[2] Op. cit., p. 38.
[3] Mantegazza, Gli amori degli uomini. Milano, 1886, p. 182.
[4] Kurz, Zwei innere Urethrotomien, gefolgt von Melancholie. Ref. der med.-chir. Rundschau. Wien, 1887, Heft xviii. p. 683.
[5] Op. cit., p. 41.

thirty years of age, who, in consequence of an injury of the occiput, was left without sexual appetite, and whose testicles were atrophied, used to talk in a joking way and quite merrily about his injury and its sad consequences. No doubt the personal character is of great importance here, as each individual shows different characteristics in the reaction following injury.

The impotent are misanthropic and distrustful. They are ever afraid that the defect of which they feel ashamed may be discovered. All this is aggravated by self-reproaches of the worst kind, for almost every one believes he has himself caused his misfortune. A reason for self-reproach is soon found, for who has not indulged in real or imaginary sexual excesses or self-abuse?

These people are so much ashamed of their deficiency that they will not acknowledge it, even to the physician, except in a most reluctant manner. It is, therefore, advisable to meet the communications of an impotent person with the required scepticism from the very beginning. This sense of shame is more intense with people of humble condition than with those of a higher social rank, who will not infrequently speak of their impotence in a joking tone, even where it might not be expected, in order thus to make you believe the contrary. I had an opportunity to observe a case where a member of the nobility was by every one considered as impotent. Under the cover of this reputation the nobleman was following up several intrigues, until at last it was discovered that this impotence was not to be relied upon.

So far we have had under consideration the influence of impotence upon the mind, and have seen that it is of no slight degree. The influence of this disease upon the body is no less important. We shall here leave out of sight circumstances which are the cause, but not the consequence, of impotence, as well as circumstances which may be the consequence of spermatorrhea.

The cessation of so important a process as the sexual function cannot occur without producing in and by itself an essential change in the individual. According to Arndt,[1] there is no disturbance of any function of a man without change in the man himself. Real impotence has a powerful influence, primarily over the mind, and secondarily over the state of health of the entire body. It is conceivable that a person who is dull or mournful and always ill-humored, who is plagued by a bad thought, must by degrees lose his appetite, suffer from indigestion, and consequently must become physically ruined; though, no doubt, one often meets, as previously said, impotent persons looking thoroughly well and healthy.

Impotence besides being a very *serious* disease is also of *frequent* occurrence. "Experience proves that the large towns especially harbor crowds of persons suffering from a diseased nervous system, who, in the different stages of life, are afflicted with sexual infirmities which throw a gloom over their existence. Youths who have scarcely stepped beyond the threshold of puberty, adults who are on the entrance or perhaps in the zenith of manhood, no less than individuals who have already reached the autumn of their lives, make up elements of these numerous groups of the sexually discontented who are suffering from diseases of the nervous system as burdensome as they are unyielding. Victims of an unequal fate, some more, some less severely wounded in the combat against untoward circumstances,—unequally

---

[1] Die Neurasthenie. Wien und Leipzig, 1885, p. 3.

furnished with chances for improvement,—all these pitiable persons are nevertheless animated with the same desire, that of being once more admitted to the full enjoyment of life and being able to found a family."[1] "It is quite incomprehensible that there should still be physicians who almost absolutely deny the existence of impotence."[2]

Every being instinctively longs for enjoyment. The desire for enjoyment is certainly justified, and only hypocrites or people with limited views of things in general can demand that man shall work and fulfil duties without the moments of gladness and gratification that are so scarce in comparison to the bitterness of life. "La nature veut que nous jouissions."[3]

Sensual love and the so-called ideal love, which grows out of it, but which is quite an impossibility where there is no sexual vigor, are foremost among the few joys and gratifications.

Those who are seeking help for their impotence are surely very pitiable subjects; they feel themselves unutterably unhappy, and, in most cases, entertain thoughts of suicide, though they seldom have the necessary courage to carry them out. A good man cannot refuse his assistance to them. It cannot be the physician's business to question whether one or another of them owes his infirmity to his own or somebody else's fault. Nor can it be expected from the medical man that he should investigate what may be the object or aim of any patient

---

[1] Rosenthal, Ueber den Einfluss von Nervenkrankheiten auf Zeugung und Sterilität. Wiener Klinik, 1880, p. 136.
[2] Lionel S. Beale, Our Morality. London, 1887, p. 34.
[3] Renan, L'abbesse de Jouarre. Paris, 1886, p. 29.

who is endeavoring to recover his virility. The physician has to keep in view this one object, that he is in the presence of an unfortunate person whom he must help if he can.

So much for the determining of my stand-point and to justify the total separation of impotence and sterility.

And now the question suggests itself, *What is impotence?* It is well-nigh impossible to give a precise answer. Maximilian v. Zeissl, for instance, gives the following definition: " Impotency is a collective idea of the various pathological details which hinder a man in the carrying out of coitus, so that the ultimate purpose, viz., that of begetting a child, is not attained in spite of sexual intercourse with a fertile woman."[1] This definition is far-reaching, because, though we may include both the impotentia coeundi and the impotentia generandi, it should yet be said that, in spite of sexual intercourse with a fertile woman, the begetting of a child " must" remain unattained.

It is much easier to give a definition of sexual virility. Sexual virility is that condition of the body, of the nerves connected with the generative organs, of the centers of these nerves, and of the genital organs themselves, which enables an individual to accomplish the sexual act with an acceptable woman always, under all circumstances, and within the limits set by nature.

This ideal condition of virility is somewhat rare with men following the customary manner of life of our days, and in any given case it will generally last for but a short time. Every deviation from this ideal condition must,

---

[1] Dr. M. v. Zeissl, Ueber die Impotenz des Mannes und ihre Behandlung. Wiener med. Blätter, 1885, Nr. 15.

indeed, be reckoned as a starting-point of impotence. The lesser deviations are not taken into account, and are generally considered as in the nature of things.

When virility is in full vigor the sight, the slightest touch, the first embrace of the desired woman must cause sexual desire and the erection necessary to the performance of the act. The individual is in the same degree approaching impotence when he requires longer preparations and longer and more intense excitation to produce the necessary sexual rousing. Of course, we leave out of question the repetition of coitus after short intervals. The reverse of this ideal condition is that state which we call total impotence, in which condition the individual can never have an erection or experience excitement, and can, therefore, never, under any circumstances, perform the act of coition. As nature never progresses by leaps and bounds, so these two forms of sexual capacity do not pass abruptly from one into the other, but between them are numberless transitory forms of impotence.

Although this essay is written for medical men only, who are conversant with the anatomy of the genitals and with the physiology of procreation, we shall devote a few pages to both Anatomy and Physiology. In the course of years and the throng of professional occupations small matters escape the memory of the practitioner, and occasionally we may read again, with profit, that with which we are well acquainted.

CHAPTER II.

ANATOMY.

We shall now give a rough sketch of the male organs of reproduction. For more minute study we refer the reader to the great number of excellent anatomical works.

The male genitalia have been divided into different sections; but as this separation contributes in nothing to the clearness of the subject, we shall simply discuss the different organs in their turn.

The *testicles* (testes, testiculi, orchides, didymi) are a pair of oviform, glandular organs. We shall first consider the coverings which protect and support them. Proceeding from the outside inward, we find first the outer skin starting from the perineum, from the inner surface of the upper thighs, from the root of the penis, and from the pubis, and forming the *scrotum* or purse. The whole of the scrotum appears asymmetrical, hanging a little lower on the left side. The slight enlargement of the cutis forming the median raphe, which runs from the perineum forward to the inner surface of the prepuce (præputium), separating externally the whole genital apparatus into two halves, and indicating the inner division on the scrotum, is not exactly in the median line, but draws somewhat to the left.

At the scrotum we find the epidermis, cutis, and tunica dartos. The *epidermis* is distinguished by the amount of pigment it contains; the *cutis*, by a strong growth of hair, sudoriparous glands, and a rich rete of lymphatics. The *tunica dartos* is a fibrous, fleshy membrane, consisting of rather strong, smooth, muscular fibers, elastic, without fat, and ligamentous in character. Still proceeding inward, we next come to the *tunica vaginalis communis*, which envelops the testicle and spermatic cord. That part which surrounds the spermatic cord is loose and spongy, containing here and there adipose tissue, and is intimately connected with the spermatic cord and the scrotum. This part of the tunica vaginalis communis consists of three layers,—an inner, an outer, and a median muscular layer. These layers or membranes are not in all parts distinctly separated from one another, because the *musculus cremaster* which separates them runs in isolated flat bundles down the spermatic cord; between the bundles the two layers run into each other. These isolated flat fasciæ of the cremaster pass in a fan-like manner downward, twining around the testicle. These muscular bundles have the power to draw the testicle upward and outward. Contraction of this muscle ensues reflexly from violent movements of the abdomen; also as the result of independent action. The testicle may, besides, be raised through the contraction of the muscular fibers in the tunica dartos.

Immediately enveloping the testicle we find the *tunica vaginalis propria*, which may be divided into two layers,—namely, the parietal membrane, which is connected with the tunica vaginalis communis, and the visceral membrane, which is united with the albuginea of the testicle and the epididymis.

As the testicles produce the sperm, they are the most important part of the male generative organs. The testicles lie side by side in the scrotum, hanging down between the thighs, below the symphysis pubis, each in its own compartment, and separated from its fellow by a median membranous partition,—the *septum scroti*. The testicles are in the abdominal cavity until the seventh month of fetal life, when they descend through the inguinal canal into the scrotum.

The left testicle hangs a little lower, this arrangement being very appropriate, as it prevents friction of the testicles in case of the sudden pressing together of the thighs.

In the testicle we have, first of all, to distinguish between the testicle proper (Henle calls it testicular gland, others call it main or chief testicle) and the epididymis.

The *testicles* are oviform in shape, flattened laterally, with the greatest diameter four to six centimeters in length, directed obliquely from above downward, forward, and outward. The weight of the testicle is fifteen to 24.5 grams, its cubic contents twelve to twenty-seven cubic centimeters, its length five centimeters, its breadth 2.5 centimeters, its thickness three centimeters. Weight and volume, length and breadth, are subject to great variations in different individuals, and fluctuate considerably even in the same individual. In spite of Henle's [1] opinion to the contrary, my experience teaches me that this fluctuation in the volume of the testicle corresponds to the amount of seminal fluid contained. It is true that the testicle does not collapse immediately after coition, but observation has convinced me that the volume increases after unusual abstinence; so that I feel sure that

---

[1] Handbuch der Anatomie. Braunschweig, 1874, p, 366.

contraction of the scrotum is here not the main cause of variation.

Having noted that the testicle is an oviform body flattened laterally, we observe further two points, the upper and the lower, and two margins (anterior and posterior) connecting them; also two flat surfaces, an inner and an outer. The superior point and the posterior margin are covered by the epididymis and the beginning of the seminal cord.

Directly investing the testicular gland itself is the *tunica albuginea*. This is a strong, fibrous membrane, brilliantly white, 0.6 millimeter thick, containing numerous ramifications of veins and small arteries, and becoming considerably thicker and more vascular toward the posterior margin. From its inner surface it sends off numerous bundles of connective tissue, and, at almost regular intervals, stronger flat transverse bundles, dividing the tissue of the testicle into numerous conical lobules, the number of which is placed by different authors at from one hundred to three hundred. Each lobule contains a great number of very fine tubules, called *spermatic canals*, or vasa seminalia. These have a volume varying, according to the degree of distention, from 0.1 to 0.2 millimeter. They inosculate with one another, and are very tortuous, so that it is difficult to disentangle them. Their number is estimated at eleven hundred.

As we have stated, the tunica albuginea becomes much thicker toward the posterior margin, this enlargement forming what is called the *corpus Highmori*. Here the seminal tubes collect, three to six inosculating, and grow less and less tortuous until they become almost straight and form the *rete vasculosum* (seu Halleri) *testis*.

From this rete start about twenty larger tubes running almost in a straight line, and passing through the tunica albuginea into the epididymis; there they form lobuli again, giving rise to that single tube with manifold convolutions which constitutes, in the main, the parenchyma of the epididymis.

Besides the seminiferous tubules, the parenchyma of the testicle contains comparatively large winding vessels with thick walls and a cellular mass, of the function of which we still know nothing positive, and which many anatomists and physiologists consider to be connective tissue. Before entering the epididymis the seminiferous tubules change in structure and become simply excretory ducts.

The *epididymis* properly considered is merely an excretory duct of the testicle. It is a body weighing generally 1.5 grams, its cubic contents being 1.78 cubic centimeters. Its upper end is globular in form and tapers off to pass into the vas deferens. The epididymis is also invested with a tough albuginea, which has the same structure as the albuginea of the testicle, but is not so thick, its thickness being only 0.04 millimeter. The inner surface of the albuginea of the epididymis also sends off septa of connective tissue into the parenchyma, dividing it into lobules, though superficially only.

The unfolded *vas epididymis* has a length of about six meters, with a diameter of about 0.44 millimeter, and gradually dilates as it approaches the vas deferens. Besides this principal duct, the epididymis contains also one to three small blind canals, the *vasa aberrantia* and the so-called *hydatis Morgagni*, which are said to be remnants of embryonic conditions.

At the lower point of the testicle the canal of the epididymis is turned directly upward in order to reach the orificium cutaneum canalis inguinalis; it is then called the *vas deferens*, and, together with the vessels and nerves running in the same direction, forms the seminal cord (plexus spermaticus seu pampiniformis).

The tortuosity of the epididymis continues into the first part of the vas deferens, but the tube becomes gradually more nearly straight, its walls at the same time increasing in thickness and extent. The total length of the seminal vessel is about fifty to sixty centimeters. According to Henle,[1] the straight part is about three millimeters in diameter, one-sixth of which is taken up by the lumen, so that the thickness of the wall is 1.5 millimeters. On this thickness depend the firmness and cylindrical form of the vas deferens.

Before the vas deferens unites with the seminal vesicle it forms the spindle-like *ampulla of Henle*. The lumen in this place becomes almost doubled in extent, the thickness of the wall increasing also. At the end of the ampulla the vas deferens grows thinner again, and has outlet in the inferior pointed end of the seminal vesicle lying at the outer part of the base of the urinary bladder, between the bladder and rectum. The end of the vas deferens forms with this pointed end of the seminal vesicle the ductus ejaculatorius.

The *seminal vesicles* are really hollow glands of a very irregular form, resembling a very knotty, somewhat flattened club. Even in the same subject the two vesiculæ seminales may differ in form and size. The length of the vesiculæ seminales varies from four to 8.5 centi-

---

[1] Op. cit., p. 382.

meters, their diameter from 0.6 to 0.7 centimeter. The superior end is blunt, usually having a hump-like protuberance, which, looked at from the outside, resembles a small hunch. The entire surface looks uneven or rough, the little hunch-like prominences corresponding to depressions on the inside.

The interior of the seminal vesicles is still more peculiar, and varies just like the exterior. The mucous membrane is of a yellowish tint, infolded, has little pits, and forms depressions and longer or shorter diverticulæ. The organ has altogether a cellular appearance. In the mucous membrane there are some peculiar glands, which, though the granular epithelium is different, have a structure similar to that of the mucous glands, but produce a secretion essentially different from mucus, as it does not congeal in acetic acid.

By the union of the vasa deferentia with the vesiculæ seminales the *ejaculatory ducts* are formed about the superior margin of the prostate gland, but with numberless variations in the share provided by the individual organs in this formation.

The parietes of the ductus ejaculatorius are about 0.4 millimeter thick, the lumen one millimeter in diameter. While the lumen runs from two to three centimeters forward and downward through the prostate, it diminishes in volume; the mucous membrane, which at first resembles that of the vesicula seminalis, losing gradually its folds and its glands, as well as its yellowish tint. The two ejaculatory canals also frequently exhibit variations with respect to form, course, convergence, and mutual contact. Even coalescence of the two ducts into one may take place. The ejaculatory canals lead into the prostatic part of the urethra near the verumontanum or colliculus seminalis, opening by circular mouths. The fact, minutely described by Henle,[1] that the muscular membrane of the ductus ejaculatorius within the prostate assumes the character of a cavernous tissue seems to me of special importance.

The *prostate*, shaped like a chestnut or flattened cone, embraces with its anterior surface the neck of the bladder and the first portion of the urethra, its posterior surface resting on the anterior wall of the rectum. Its texture is firm, the borders rounded off. The superior border or margin which surrounds the bladder is broader, slightly bent in the middle, while the inferior margin tapers off. The greatest diameter of the prostata measures thirty-two to forty-five millimeters; from the base to the point it is twenty-five to thirty-five millimeters; its thickness, fourteen to twenty-two millimeters. Its weight is estimated at seventeen to 18.5 grams.

According to Henle, the prostate comprises three different organs, or, rather, structures,—a number of racemose glands, the glandula prostatica; a closing muscle of the bladder, composed of smooth muscular tissue, the sphincter vesicæ internus; and a transversely striped closing muscle of the bladder, the sphincter vesicæ externus. Besides these, we have to notice in the wall of the ductus ejaculatorii, of the sinus prostaticus, of the urethra, and of the colliculus seminalis, the peculiar structure which sends off shoots into the substance of the prostata and also the exterior coat of the glandular portion, together with the separating walls or septa starting from it.

The main substance of the prostata is the real gland,

---

[1] Op. cit., p. 388.

which does not reach complete development until after puberty, as the glandular ducts and vesiculæ develop greatly at the expense of the substance of the connective tissue, which predominates before puberty. They give to the whole gland a yellowish-red tint and a spongy appearance.

The relative proportion of the muscular fibers and the glandular substance in the prostate varies considerably in different individuals. In one subject the glandular spaces may be predominant, and in another their contractible coatings, so that in one person the secreting function of the prostate may predominate, and with another the motory.[1] Unfortunately, we have, as yet, no observations to determine what influence this difference in the structure of the prostata has on the sexual life.

The excretory ducts within the prostata unite into an indefinite number of stems, which open into the urethra at the colliculus seminalis anterior to the mouths of the ductus ejaculatorii. Two of the largest stems open almost symmetrically, side by side, quite close to the openings of the ductus ejaculatorii; the others, from seven to fifteen in number, open rather more in front, asymmetrically and with variations.

The *secretion of the prostata* is of the nature of mucus, but with acetic acid congeals but slightly.

We now pass to a short description of the real organ of copulation, the penis, which, with its corpora cavernosa, is perforated by the urethra.

The *urethra*, which runs from the neck of the bladder

---

[1] Rüdinger, Zur Anatomie der Prostata, des Uterus masculinus und der Ductus ejaculatorii. München, 1883, p. 4.

to the exterior orifice of the penis, is divided into three portions,—the pars prostatica, pars membranacea, and pars cavernosa. The course of the urethra resembles the letter S,—*i.e.*, it has two bends or curves, of which the posterior retains its shape even during erection. The length of the urethra varies very much, and is from fourteen to twenty-two centimeters, the pars prostatica = two to 2.8 centimeters, the pars membranacea = 1.50 to 2.50 centimeters, and the pars cavernosa = 10.5 to 16.7 centimeters. There is a similar variation in the lumen of the urethra. It is narrowest in the pars membranacea, and of varying width, but least extensible, at the orificium externum.

In the pars prostatica the lower wall stands out, forming the *colliculus seminalis* (caput gallinaginis, verumontanum, crista urethralis). This is the most important part of the urethra, with regard to the subject we are treating, as it is the seat of many diseases. The caput gallinaginis begins, according to Henle, at the urethral mouth of the bladder, with two longitudinal folds, converging toward the median level space; along with these two there is occasionally a third, the median fold. The caput gallinaginis may begin with a greater number of smaller folds. This crest reaches its greatest extent, in height and breadth, at about the middle of the pars prostatica, immediately before (under) the sharp bend; it then decreases again even less abruptly, its transverse diameter diminishing at the same time. The anterior end extends, in the form of a narrow ridge, far into the pars membranacea, and often divides, toward the end, into fork-like branches at an acute angle.

This description is very definite, indeed, and indicates clearly the great variations we meet in the structure of the crista urethralis. These differences become still more numerous in consequence of disease or the frequent use of instruments. Age also may have a great influence on the formation of this organ, so that variations are met with at every autopsy as well as at every endoscopic examination. The same thing may be said of the measurements of the breadth and height, stated to be about three millimeters.

The mucous membrane of the crista gallinaginis is laid in small creases, which open out during erection.

Besides the above-described openings of the ductus ejaculatorii and of the excretory ducts of the prostate, we find at the anterior slope of the colliculus seminalis a slender follicle without outlet, the *sinus prostaticus* (Morgagni, also utriculus prostaticus or vesicula prostatica). It can scarcely be determined what function, if any, this structure has. At any rate, the statement of Rüdinger,[1] that the uterus masculinus has remained capable of contraction in a very high degree, in virtue of the smooth muscular fibers, which can be demonstrated in all parts, may perhaps seem to justify the assumption that it performs some functional, possibly a secretory, part or rôle.

The *pars membranacea* (s. carnosa, s. isthmus) *urethræ* is that part of the urethra which, leaving the prostata, penetrates into the diaphragma urogenitale to enter the corpus cavernosum urethræ at the inferior surface of the diaphragm; from this point on it is called the pars cavernosa.

Next to the posterior border of the diaphragm, between the layers of the musculus transversus perinæi

---

[1] Loc. cit.

profundus, lie *Cowper's glands*, belonging to the racemose variety. These are two lobulated bodies, resembling a mulberry, spherical, sometimes pressed flat, and measuring from four to nine millimeters in diameter. Their excretory ducts, three to six centimeters long, converge and have their outlets close together in the urethra, at the end of the bulbous and somewhat dilated part of the pars cavernosa.

The mucous membrane of the *cavernous portion* of the urethra is in longitudinal folds. Besides the outlets of Cowper's glands, it contains the very fine *glands of Littré* (0.1 millimeter average diameter) and the very small *lacunæ Morgagni*, dot-like in appearance. The lumen of the urethra is dilated at both ends of the cavernous portion, corresponding to the bulbous part and the fossa navicularis.

The entire urethra is lined with cylindrical epithelium, which in the fossa navicularis, and sometimes a little before, changes into pavement epithelium. Within the region of the pavement epithelium there are papillæ sometimes 0.22 millimeter in height and of diverse forms. Vajda[1] declares that he has discovered vascular papillæ of sundry sizes and shapes in the whole mucous membrane of the urethra, nearly as far as the pars bulbosa; and that the pavement epithelium of the fossa navicularis extends over the whole surface of the urethra.

The wall of the urethra consists of the mucous membrane, to which is annexed a layer of areolar tissue, the meshes of the latter being stretched in the longitudinal direction of the urethra. This areolar layer is, in the

---

[1] Beiträge zur Anatomie des männlichen Urogenital-Apparates. Wien, 1887.

prostatic part, the membranous part, and the first portion of the cavernous part, enclosed by a layer of smooth muscular texture, with which many elastic fibers are interwoven.

The areolar tissue, which constitutes the areolar layers of the ductus ejaculatorii, the pars prostatica urethræ, and the pars membranacea urethræ, is called by Henle *compressible areolar tissue*, in contrast with the erectile areolar tissue, of which the corpora cavernosa urethræ et penis consist.

The pars cavernosa urethræ is enveloped in a cylinder of areolar tissue, which, toward the posterior end, gradually thickens to a club-shape and forms the so-called bulbus urethræ; while the anterior part suddenly spreads out, covers the anterior ends of the corpora cavernosa penis, and thus forms the glans penis. Each of the anterior ends runs off into a blunt point, which is covered by the anterior expansion of the corpora cavernosa urethræ.

The *corpora cavernosa penis* are a pair of bodies of cylindrical shape, slightly flattened on the inside. They come in contact in the even median surface, while their posterior ends, the so-called roots, diverge and fasten themselves on the inner surface of the lower border of the inferior branch of the pubis. The superior and lateral surfaces of the corpora cavernosa can be felt through the outer skin, while the inferior surfaces in their contact form the urethral furrow for the reception of the corpus cavernosum urethræ.

We must be brief in the description of the corpora cavernosa, as details would lead us too far. Each corpus cavernosum has a ligamentous, brilliant white tegument, consisting of connective tissue and elastic fibers, in which there are a few very sinuous blood-vessels. This cover, called *albuginea of the corpora cavernosa*, is about two millimeters thick when the member is flaccid, but grows much thinner when the corpora cavernosa are filling.

From this albuginea proceed into the interior of the corpora cavernosa transverse vascular bundles of connective tissue, consisting of elastic filaments and smooth muscular fibers, and parietes or septa, with small interstices between. Thus is formed the spongy texture of the corpora cavernosa. These small interspaces are coated with vein epithelium; all are interconnected by emissariæ.

Henle asserts that these interspaces are vascular plexuses between the ends of arteries and the beginnings of veins, as neutral in character as the capillary rete of other tissues. They may be considered capillaries which have dilated and run together at the cost of the intermediate tissue, partly through atrophy of the latter, and which have reduced the intermediate tissue to a number of transverse bands and leaf-like septa, in which run some supplying vessels as well as ordinary capillaries of the usual diameter.

We omit as unnecessary a description of the cutis of the prepuce, the frenulum, and, in a word, of the exterior of the penis. Every physician is aware of the individual differences in the volume of the entire penis, of the length and form of the prepuce.

The inner surface of the prepuce is devoid of hair, smooth, and shining. The surface of the glans when the member is flaccid is slightly furrowed, and consequently dull in appearance; during erection it becomes even and shining. There are numerous sebaceous glands on the arched surface of the glans, on the prepuce, and around the frenulum; moreover, there are many papillæ differing in number and size. Sometimes single papillæ are found also on the inner surface of the prepuce where it joins the corona.

CHAPTER III.

PHYSIOLOGY OF THE SEXUAL ACT.

We shall now consider the male sexual functions, leaving fecundation out of the question, and shall give our attention to a short sketch of the physiology of coitus alone, without mentioning the processes which cause fecundation. Besides, in the following chapters we shall pay due attention to the physiology of our subject.

It is generally asserted that, with regard to procreation, nature has imposed on the woman all the burdens, and reserved only pleasure for the man. This is so, indeed, if we take into consideration that the woman, after conception, has to carry and nourish the fruit in her body for nine months and then undergo the labor of parturition. So far as coitus is concerned, however, in and for itself, the greater part devolves upon the man, and, moreover, in comparison with the woman, he is at a great disadvantage. I shall not dwell longer on this subject, as, strictly considered, it does not enter into the framework of our purpose, and I shall make the man alone the object of my discussions.

Sexual Maturity.—In order to perform normal coitus the individual must be in possession of all the qualities necessary; in the first place, he must have attained *puberty*. In our climate males reach puberty, on an average, at the age of seventeen. Puberty announces itself by various exterior signs, the most striking being the alteration of the voice, which grows deeper and sounds rough and broken during the period of sexual development known as the age of puberty. This deepening of the voice is caused by a certain series of changes in the larynx: the processes vocales become cartilaginous, the larynx larger and protruding, the vocal cords lengthened. Furthermore, the bones and muscles grow stronger, the lungs larger; the pubic region becomes covered with hair. But the greatest changes occur in the genitals, the testicles enlarging and beginning to secrete. The tissue of the penis, capable of enlarging, develops disproportionately, and the prepuce loosens from the glans. The sexual impulse awakens, and, if not satisfied, results in pollutions.

This transition of the child into a pubescent man requires about two years for its accomplishment. Thus, generally speaking, the young man would be nubile at the age of seventeen.

Here I must assert my opinion, in opposition to others, Roubaud, for instance, who says that spermatozoa are never found in the semen of youths under eighteen years of age. I have repeatedly discovered perfectly developed spermatozoa in the semen of Frenchmen, Italians, Croatians, and Hungarians hardly sixteen years old. I must remark, however, that most of these were youths who sought for sexual gratification prematurely.

About eleven years ago I performed the autopsy of a sixteen-and-a-half-year-old shepherd, who had been accidentally drowned, and whom his comrades had declared to be an onanist. I found in the testicles, as well as in the excretory ducts, spermatozoa in every grade of development. I also found a great quantity of spermatozoa in the semen of the second pollution of a Croa-

PHYSIOLOGY OF THE SEXUAL ACT. 47

tian peasant boy, who was not quite sixteen years old, who was not an onanist, and who had not had sexual intercourse with the opposite sex.

The individual retains his power to perform the sexual functions during a greater or less period of his life. There are people who, in their fiftieth year, become sexually useless in a quite normal manner, according to constitution, temperament, and habits. On the other hand, it is impossible to state any age at which there have not been or may not be men sexually capable. The procreative power is, however, most likely to be extinguished after the age of sixty, whilst the capacity for intercourse is certainly preserved much later. Girault found that the spermatozoa change after the fifty-fifth year, the heads growing larger and the tails shorter; this alteration certainly not contributing to their power of movement.

Sexual Orgasm.—The copulative power requires not only that the individual be virile and his sexual instinct unextinguished, but also that he be capable of having the sexual orgasm (libido sexualis), which is a combination of centrally or peripherally roused fancies and pleasurable sensations associated therewith. The libido sexualis is very frequently in itself a peripheral excitement, and in the sexually virile the center of erection acts promptly through the afferent and efferent nerves. We shall see in our future considerations that there are many cases where the impotence is merely a consequence of complete or incomplete absence of sexual excitement. We shall see that anything capable of distracting the sexual excitement at the given moment, to divert the run of ideas from the sexual track, is also capable of causing sexual impotence, be it only for the moment.

What is it, then, speaking generally, that can cause the sexual orgasm? If the human male is left to himself and nothing comes to disturb the natural course, the first sexual rousing will not occur until he has reached the state of full sexual maturity; when the testicles, spermatic ducts, and vesiculæ seminales are filled with sperm. This first sexual excitement would occur even if the individual should have no idea of sexual things, which, probably, would be exceptional. This is a proof that the accumulation of sperm in the seminal organs may and must occasion sexual excitement quite independently of the will of the individual. On the other hand, we see that sexual excitement takes place with persons whose seminal organs are anything but full of sperm. It is evident that here the sexual excitement comes about through mental impressions. The center for mental impressions is in the cortex cerebri. Therefore sexual excitement may be caused through the medium of the cortex cerebri, and this is generally the case.

We see individuals whose sexual sense is deadened in such a wise that the psychical impressions alone can no longer bring about sexual rousing. Such people have generally no sperma in their spermatic organs to cause the rousing, and yet they are known to accomplish coition. It is well known that these persons are in the habit of putting themselves into a state of sexual excitement by irritating the exterior part of the organ, and sometimes have recourse to the most loathsome manipulations.

Indeed, we see that the most diverse irritations, practised on the outer and the inner nerves of the sexual organs, cause libido sexualis. Thus, for instance, incidental friction of the genitals by too tight garments, the introduction of a sound or a catheter, an inflammatory or catarrhal condition of the urethra, pressure of a full bladder or of the rectum, anal fissures or hemorrhoidal ulcers, irritation by worms or by urine containing some medicinal or certain alimentary substances,—all these may cause hyperemia or secondary libido sexualis.

Finally, it is known that from the most diverse organs sexual erethism can be aroused; above all, from the organs of the senses, particularly the organs of sight, touch, and smell. It must be observed, however, that these cause sexual excitement indirectly only, by means of the central organ, the cortex cerebri. The sight of a beautiful woman, the touch of certain parts of the body of a woman, the agreeable odor of a woman, a lascivious picture, all are apt to bring about sexual excitement, but only in so far as such impressions on the senses give rise to some idea or recollection in the central organ.

Sexual excitement may arise, therefore, in three distinct ways: (1) reflexly and naturally, through the accumulation of sperm in the seminal organs, in which case there is no intervention of the cortex cerebri, the seat of sensation and ideation; (2) psychically,—the most frequent way,—through ideas, consequently through the activity of the cortex; and (3) unnaturally, by means of direct excitation of the sexual organs.

I leave out of question here single and rare cases, in which it is pretended that the sight of an object in no way related to sexual things, odors certainly not coming from a woman, the touch of objects not in any relation with woman, the eating of certain food not physically aphrodisiac, or even impressions on the sense of hearing, have produced sexual excitement. Observations of this nature, if they do not belong to the province of fable or are not based upon error, can be made only upon individuals psychically abnormal, and in most cases can be traced to fancies or recollections.

Thus we see that in sexual matters the *cortex cerebri and the sexual organs are in mutually dependent relations.* Ideas and desires which originate in the cortex cerebri act upon the sexual organs through the medium of other centers situated in the lumbar region; on the other hand, certain occurrences in the sexual organs, principally such as accumulation of sperm, create images and ideas of a sexual nature in the cortex cerebri, which may seem unaccountable to an inexperienced person, but which result in libido sexualis.

Seat of the Sexual Instinct.—It has not yet been discovered which part of the cortical substance is the *seat of the sexual instinct.* Besides the center of excitement there must be a *center of inhibition,* which, with people who have learned to control themselves, is developed in a higher degree, and puts a kind of check on the libido sexualis. It is probable that this inhibitory center does not exist in animals at all, and is but slightly developed in people on a low moral plane. However, this inhibitory center, often very beneficent, has also its disadvantageous side. In the numerous cases of sexual neurasthenia called by various names, the cause lies in the untimely interference of this inhibitory center.

The *sexual capacity* of an individual depends mostly on the facility with which he can be brought into a state of sexual excitement. Lively and excitable men, who are easily and on every occasion thrown into sexual excitement, are more prone to excesses in venery, and, all other conditions being equal, accomplish more in this respect than phlegmatic, cold natures, with whom it requires a concurrence of several circumstances in order to produce sexual erethism. This fact explains also the well-known sensuality of artists, to which attention was called in the Graef case at Berlin, and which derived a more conclusive proof from the prudish protest of the artists themselves.

Erection.—Sexual irritation causes the *erection* of the virile member. An erection may, however, take place without the excitement by reflex action; but such an erection is not sufficient for coitus without the addition of sexual erethismus. We apply the term erection to that physiological process which puts the virile member into the condition that enables it to make its way into the vagina.

We shall now direct our attention to what takes place in erection. In the chapter on Anatomy we have described the structure of the penis, and we have seen that, from the tunica albuginea of the three erectile bodies, vascular transverse fasciae and septa run into the inner part of the corpora cavernosa, leaving small interstices, thus converting the corpora cavernosa into spongy bodies. These small hollow interspaces of the three corpora cavernosa are coated with endothelium resembling that of the veins, and are consequently venous spaces. Numerous emissaries keep all the corpora cavernosa in communication with one another, and open out into the vena dorsalis and the vena profunda penis. In the base of the penis there are the arteriae helicinae, which are wound in the shape of a ram's horn, in order that they may yield to the changes of volume in the erectile tissue. It is now clearly demonstrated that erection is caused by a filling of these spaces with blood, but the entire process of erection in man is nevertheless far from being explained. The researches on this subject by Kölliker, Rouget, Langer, Eckhard, Goltz, Lovén, and Frey are highly meritorious, it must be admitted, but the *mechanism of erection* has still its mystery.

We know that erection is the result of an increased influx of blood into the areolar tissue of the corpora cavernosa, together with a decreased outflow of blood from the same bodies; but we are far from understanding the cause of that afflux, often quite sudden, and of that checking of the outflow. At first an attempt was made to explain the process by the macroscopic anatomical relations. It was supposed that either the outer transversely striped muscles or the inner smooth muscles exercise a pressure on the efferent veins. Kölliker's opinion was that the relaxation of the smooth muscles caused erection. Opposed to this opinion is the theory that the smooth muscles in the walls of the cavernous spaces do not possess alone sufficient strength and energy to influence erection to such a degree; for when the nervus dorsalis penis, which innervates the transverse fibrous bands of the areolar tissue of the penis, has been severed the erectile power is reduced. Eckhard has even painlessly irritated the nervous dorsalis penis of dogs under chloroform without producing an erection; nor has an erection ensued after the excitation of the central end of the severed nervus pudendus communis of a dog.[1] If it were not for these experiments, much might be said for the opinion of Kölliker; for, if these muscles were powerful enough, their contraction

---

[1] Eckhard, Ueber den Verlauf der Nerven erigentes innerhalb des Rückenmarks und Gehirns. Beiträge zur Anatomie und Physiologie, Band vii. Heft 1, p. 70.

PHYSIOLOGY OF THE SEXUAL ACT. 53

would certainly prevent an erection. On the other hand, the fact that warmth causes a dilatation and cold a contraction of the corpora cavernosa speaks with some force for the influence of the smooth muscles. However, this is not sufficient for an explanation of erection, and consequently other theories regarding the physiological process of erection have always been sought. Thus, Eckhard obtained, through his experiments upon animals, the following results: Erection can be caused in rabbits by faradization of the lumbar and cervical regions of the cord, the pons Varolii, and the peduncles of the cerebrum, while irritation of the cerebellum alone has no such effect.[1] Eckhard is therefore of the opinion that the seat of erection is in the cerebrum, and that the nerves which bring about erection, starting from the cerebrum, run downward in the medulla spinalis.

Goltz, in his experiments upon dogs, cats, and rabbits, found that erection can be caused by excitation of the glans, bladder, and rectum, even after the spinal cord has been severed at the upper border of the lumbar region, and that it is also possible to cause ejaculation of sperma after destruction of the lumbar region of the cord, even though the capability of erection has been entirely extinguished.[2]

The result of these experiments on animals corresponds with clinical observations in cases of injuries and diseases of the spinal cord. Further experiments have shown that in the cerebrum and in the upper portion of the spinal cord there must be inhibitory nerves affecting

---

[1] Eckhard, op. cit., p. 77.
[2] Goltz, Ueber die Functionen des Lendenmarks des Hundes. Pflüger's Archiv für die gesammte Physiologie, Band viii. Heft 8 u. 9, p. 464.

erection; because, after section of the cord at the upper border of the lumbar region, erection can be caused more easily and more vigorously by means of electric irritation.

We have seen that erection is caused by afflux of arterial blood and an obstructed outflow of venous blood. Erection cannot, however, be brought about by mere compression or ligation of the veins. That an increased flow of arterial blood takes place is proved by the lowering of the blood-pressure in the neighboring inguinal arteries, and even as far as the arteria cruralis, as shown by the manometer. How this increased arterial afflux is caused is not yet explained. Certainly the acceleration of the heart's action which always takes place during sexual erethism can have but a very slight influence. Possibly, the arteries of the penis, in consequence of the sexual erethism, become dilated, and thus admit more blood. The question is, Has this erethism a paralyzing influence on the muscles of the walls of the vessels, or is it an invigorating effect, causing the contractions to become more energetic and more frequent, and the arteries to pump, so to speak, a greater quantity of blood into the mesh-like spaces of the penis?

Goltz endeavors to explain the act in this way: "I share the opinion of those who compare the connection of the nervi erigentes with the penis to that of the vagus with the heart, or the chorda tympani with the vessels of the glandula submaxillaris. The progress of blood through the penis is considerably hindered during the time of physiological rest; for then the small arteries of the penis and other vascular spaces are in a state of moderate contraction. Very likely this state of tension or firmness of the vessels of the penis is maintained by the action of the small ganglia whose presence in the penis has been proven by Lovén. The pressure of the ingoing blood dilates the arteries; when the resistance is removed the blood flows more freely through the areolar tissue into the erectile bodies, and puts them into a state of turgescence. Being, then, with Lovén, inclined to consider the peripheral ganglia the tonic center of the vessels of the penis, we must suppose that the nervi erigentes during their activity paralyze or check these ganglia in the same way that the vagus checks the activity of the ganglia of the heart."[1] Goltz saw this opinion confirmed by the fact that he succeeded in proving experimentally that reflex erection of the penis could be prevented by a more intense irritation of the nervus ischiadicus.[2]

Such experiments, if continued, will no doubt at some future time result in a complete explanation of the mechanism of erection, though for the present the practitioner can obtain but unimportant benefits from the above researches.

Experience teaches us that erection can be either caused or checked by different impressions from the most varied parts of the body. It is certain, however, that the cerebrum is the place of origin of the sensations of sexual excitement. With this higher center is connected, by intercentral nerve-channels, an inferior, mechanical reflex center, which has its seat in the lumbar region of the cord, and governs the performance of the

---

[1] Goltz, op. cit., p. 466.
[2] Compare: Prof. Dr. A. Spina (Prague): Experimentelle Beiträge zu der Lehre von der Erection und Ejaculation. Wiener med. Blätter, 1897, No. 10–13.

act of copulation.[1] It is probable that from the spinal cord issue some special nerves which straighten the erectile vessels or else diminish their extensibility.[2]

When erection is accomplished, the penis is in a condition that makes penetration possible. We shall now direct our attention to the process of coitus. Roubaud[3] describes it as follows: "The moment the membrum virile enters the vestibulum, the glans penis rubs itself first against the glandula clitoridis, which is situated at the orifice of the sexual canal, and can yield and bend in consequence of its place and of the angle it forms. After this first excitement of the two centers of sensation, the glans penis glides over the borders of the two bulbi; the collum and corpus penis are enveloped by the prominent parts of the bulbi, while the glans, having penetrated deeper, has come into contact with the fine and delicate surface of the mucous membrane of the vagina, which itself is elastic by virtue of the erectile tissue between the membranes. This elasticity, which enables the vagina to surround the body of the penis closely, also increases the turgescence and, consequently, the sensibility of the clitoris, by conducting the blood, driven out of the vessels of the vaginal walls, to the bulbis and clitoris. On the other hand, the turgescence and sensibility of the glans penis are heightened by the compressive action of the tissue of the vagina and of the two bulbi in the vestibulum, which tissue becomes more and more turgescent. Moreover, the clitoris, pressed downward by the anterior portion of the musculus com-

---

[1] Goltz, op. cit., p. 473.
[2] Eckhard, op. cit., p. 80.
[3] Traité de l'impuissance et de la sterilité. Paris, 1876.

pressor, meets the dorsal surface of the glans and corpus penis, and causes friction. This mutual friction at every motion during copulation affects both individuals, and leads finally to that high degree of orgasmus which induces, on the one hand, the ejaculation, and, on the other, the reception of the seminal fluid in the open orifice of the neck of the uterus.

"The question has been asked—and many earnest investigators have tried to answer it—whether the greater share of pleasure falls to the man or the woman in these moments of the greatest intoxication. This question has received the most contradictory explanations, like all questions that rest on essentially different and varying bases. Indeed, taking into consideration all the circumstances which influence the sexual sense, I do not believe that it is possible to solve the problem *a priori*. If we consider how much the sexual sense is influenced by the temperament, the constitution, and a great many other circumstances, special as well as general, we shall be convinced that this question is far from being solved—nay, is insoluble; and this is so true that we meet with difficulties even if we want to draw a true and perfect picture of coitus: while with one individual the sensation of voluptuousness manifests itself merely by a slight quivering, it will reach, with another, the climax of exaltation, moral as well as physical. Between these two extremes there are innumerable gradations. Increased rapidity of the circulation of the blood, violent pulsation of the arteries, the retention of venous blood in the vessels by the contraction of the muscles, increases the general warmth of the body. The stagnation of the venous blood becomes still more decided in the brain, on account of the contraction of the muscles of the neck and because the head is bent backward. All this causes a transitory congestion of the brain, during which reason and all the mental faculties are lost. The eyes, reddened by the injection of the conjunctiva, become staring, the look becoming unsteady; or, as is more generally the case, the eyes close convulsively to avoid contact with light. The respiration is gasping with some, whilst with others it is interrupted by a spasmodic contraction of the larynx; and the air, compressed for a time, forces an issue at last, in connection with disconnected and incomprehensible words. The nerve-centers, as was just said, give confused impulses: movement and sensation are in indescribable disorder: the limbs are seized by convulsions, sometimes also by cramp; they move in all directions, or stretch themselves and grow as stiff as bars of iron; the jaws, pressed together, make the teeth gnash; and some individuals go so far in their erotic delirium as to bite to bleeding a shoulder that the partner may have exposed incautiously.

"This frantic condition, this epilepsy, this delirium, generally lasts only a short time, but sufficiently long to exhaust completely the strength of the organism, especially with the man, where this hyperexcitation ends with a more or less abundant loss of sperm. Then follows a state of exhaustion, the intensity of which is in proportion to that of the previous excitement. The sudden weariness, the general weakness, the inclination to sleep, which take hold of the man after copulation. are to be ascribed partly to the expulsion of sperma; because the woman, however energetically she may have co-operated in the act, feels only a passing lassitude, which is far less than the weariness of the man, and which allows her, after a much shorter lapse of time, to repeat the act. 'Triste est omne animal post coitum, præter mulierem gallumque,' says Galien, and this axiom is, in substance, correct as far as the human race is concerned.

"From this moment the further function of generation is not attended with consciousness; the part of the man is at an end, and the woman begins now to bring her capabilities into action."

I have quoted entire this classic description of the act of copulation, although I cannot agree with all the details expressed. The mechanical act being so universally known, further comment on it is quite superfluous; indeed, it would be difficult to say anything new or differing from that already related by Roubaud. It would seem, however, incorrect to assume that the hyperemia of the clitoris should depend on anemia of the vaginal walls caused by the friction of the penis. If the act is normal—that is, if there is not too great a disproportion in the size of the respective genital parts—there cannot possibly be any such pressure exerted on the vaginal walls as to cause an anemia, on account of the great elasticity of the tissue. This observation is merely made incidentally, as this point does not enter into a consideration of our subject.

Ejaculation.—Coitus ends, for the man, with ejaculation, after which the erection gives way in most cases. Occasional exceptions occur, however, though this is contrary to the usual opinion. With particularly vigorous men the relaxation of the member is not so rapid, and an immediate repetition of the act is possible, and is actually practised by men who do not care particularly about cleanliness. If, after ejaculation, the man continues the movements of coitus, it is done generally out of gallantry toward the partner; all that the man wants after ejaculation is rest.

Of coitus it may be said with full justice, "Finis coronat opus," for ejaculation, the end of the act, is also its most essential and decisive part. The process in ejaculation has been observed to some degree, but only in animals. The effect of the friction on the sensitive glans is communicated to the musculi ischiocavernosi, called by Visale erectores penis, and to the musculi bulbocavernosi, which contract and convey more and more blood to the glans. This superabundance of blood invades the prostate and the neck of the bladder also, thus completely closing the bladder. By the continued irritation the contents of the ductus ejaculatorii, the prostata, and the seminal vessels are expelled, these fluids combined constituting the ejaculated sperma. It is probable that the muscles of the prostata and of the pars membranacea contract first, and then are immediately followed by the musculi ischiocavernosi and bulbocavernosi, which are much stronger and constitute the principal element in ejaculation.

The center for ejaculation has been proven, by Budge, to be at the level of the fourth lumbar vertebra.

With the contraction of the prostata begins an intensely voluptuous sensation, but, in case the movements of coitus cease here, the ejaculation can be retained, which possibility is frequently taken advantage of by persons expert in coition.

Ejaculation takes place also in nocturnal pollutions. Here the excitation, caused by other conditions, leads also to a contraction of the muscles, and if the sleeper awakes before the musculi ischiocavernosi and bulbocavernosi are contracted, he may prevent the issue of semen.

Ejaculation is accompanied by an intensely pleasurable sensation, which is dependent on the exciting of the sensitive terminal branches, or rami, of the nervus pudendus running out into the glans. A number of these rami end in the genital nerve-bulbs. The bulbi, irritated by the friction against the stretched surface of the glans, covered with delicate skin, lead this irritation toward the center.

Ejaculation, as we have seen, is the physiological process by means of which the sperm is sent forth toward the mouth of the uterus.

The Semen.—We must now turn our attention finally to the product of the male genital organs,—viz., the semen.

Hippokrates considered that the whole body was employed in the production of the semen. We know that this fluid is a mixture of the products of sundry glands within the sexual apparatus of the male, and that it consists of the secretions of the testicles, the epididymes, the vasa deferentia, the seminal vesicles, Cowper's glands, and the prostata.

Ebner has very thoroughly studied the process of the preparation of the sperma in sections of the testicles of rats, and found that it proceeds from special cells, ending in lobules, and advancing like columns toward the inner portion of the canal. The cells are called *spermatoblasts*.

Within a seminal canal are to be found, at the same time, all the different degrees of spermatic formation; all the forms of development repeat themselves about twice within the space of from ten to fourteen millimeters.

Landois, who worked at the same time and independently of Ebner, obtained the same result. He calls these spermatoblasts spermatic ears or spikes.[1]

Many different views prevail with regard to details in the preparation or production of sperma, and this subject is far from having a satisfactory explanation. Quite new are some results to which von Bardeleben came in the course of his researches, which, if confirmed, will revolutionize all our views upon the subject.

The *seminal corpuscles* are the product of the testicles. Every corpuscle shows three characteristic portions,—viz., a head, a middle piece, and a tail. The head is stained by carmin, and therefore is to be considered the nucleus. The middle piece has the form of a delicate little rod, or of a cone, and connects the head with the thread-like tail. This middle piece is stained by iodin more than is the head. The whole semen corpuscle measures 0.051 millimeter, of which the head forms 0.005, the middle piece 0.006. and the tail 0.04.

The tail is the principal motor organ of these spermatozoa, the mechanism of motion being explained differently by different investigators. The energy of motion of the individual corpuscles is variable. Hensen[2] estimates the time of half an oscillation at at least one-quarter of a second during the continuance of full vital power. The rapidity of the movement forward is between 1.2 and 2.7 millimeters, or, according to others, 3.6 millimeters per minute.

The semen corpuscles, formed in the testicles, remain

---

[1] Landois, Lehrbuch der Physiologie. Wien und Leipzig, 1893, p. 1012.

[2] Physiologie der Zeugung. Hermann's Handbuch der Physiologie, Band vi. Theil 2, p. 92.

there until they are discharged in the usual manner of seminal expulsion. It is hard to believe that spermatozoids once formed are reabsorbed. There are, it is true, some isolated observations which seem to support the idea; as, for instance, Schweiger observed spermatozoids of a young ram undergoing granular decay; this and Kehrer's experiments in applying ligatures to the vas deferens of rabbits are well-known cases. Finally, as a curiosity, may be mentioned the reabsorption of urine in the bladder, described in the *British Medical Journal*.[1]

Recently the question of the absorptive power of the bladder has been ventilated *cum studio*, but somewhat, also, *cum ira*. While Alapy[2] comes to the conclusion that absolutely nothing can be absorbed by the bladder itself, Hottinger's[3] experiments seem to prove the contrary. If we remember that a limited absorption can be accomplished even by the skin, it is hard to understand why the intact bladder should not, under favorable circumstances, be able to absorb to some extent.

Dr. Gerota, of Bucharest,[4] showed by a series of microscopic plates that, while there was imbibition on the part of the mucous lining of the bladder, there was no real active absorption.

The philosophical discussions of Gosselin, Haller, and others show only that these men had no correct conception of the sexual functions of the man.

---

[1] Black, On the Functional Diseases of the Urinary and Reproductive Organs. London, 1875, p. 157.

[2] Centralblatt für die Krankheiten der Harn und Sexual Organe, 1896, p. 328.

[3] Ibidem, p. 333.

[4] Medical Record, September 18, 1897.

To me this theory of reabsorption of semen is comparable to the notion that a cold can be caused by the absorption of perspiration, which Hebra, in his lectures, ridiculed whenever an occasion arose. Hensen[1] says, in referring to the above, "We require better proofs to establish a normal decay. Decaying of elements in the sperma would of necessity be so unprofitable, even dangerous, for the race that some contrivance to excrete them would long since have formed." Hensen thinks that the sperma, being slowly formed, will gradually be driven or pushed out of the ductus ejaculatorius, in case pollutions alone do not favor the continued renewal of sperma. Hensen, therefore, supposes a physiological spermatorrhea. This also is an assumption which would likewise require demonstration. I, myself, during the years 1884 to 1888, made observations on this subject, and, although they may not constitute a conclusive argument against Hensen's assumption, I will state them as an incentive to further investigations.

During two years I have, with special care, repeatedly examined the urine and urethral mucus of two perfectly healthy and vigorous men, aged twenty-nine and thirty-four, who were absolutely continent and had had no seminal emission; and I have never found even a trace of spermatozoids. The younger of these men, a clergyman, submitted to this examination on account of a friendship of long standing between us and his love of science. Before the first discharge of urine in the morning, for seventy-five successive days, I examined carefully, with the microscope, the urethral mucus, which was small in quantity, and then every discharge

---

[1] Op. cit., p. 93.

of urine; but never found as much as a trace of a spermatozoid. I lay especial stress upon this as opposed to the assertions of Black,[1] who states that he had frequently found seminal filaments in the urethral mucus after alvine evacuations in healthy men.

The above-mentioned man, thirty-four years of age, had been married, but his wife died of phthisis six years before I began my observations upon him. He had lost his two children by diphtheria in the same year. The unfortunate man grew melancholy and for some time felt no sexual desire. Later on he had pollutions and violent sexual desires; but he overcame them, and after four years of absolute continence he seldom had erections in the mornings, and thought no longer of sexual passions. He asserted that formerly he had been vigorous in coitus and had indulged daily.

The twenty-nine-year-old man had masturbated at wide intervals from the age of sixteen to twenty-three. He struggled against this "sinful practice," as he called it, with all the power of his exceptionally strong character. He finally mastered his passion. For two years his occasional pollutions grew less frequent, and for three years past he has had no emission of sperm in any way; sexual desire visits him seldom, though he has erections every morning.

Still more characteristic is the following observation. A professor at a university, thirty-nine years old, and of large and powerful build, has been married for fourteen years, enjoying good health. His wife, often sickly, disinclined to sexual intercourse, is absent for months; hence a forced continence for long periods. As soon as

---

[1] Op. cit., p. 63.

this goes beyond a month there appears a slight but painful swelling of the testicles. He finds himself compelled to indulge in sexual congress at any cost, during which he emits unusually large quantities of sperm, and the swelling and pain in the testicles vanish as if by magic.

I claim no conclusive credit for the above three observations; but they, together with a number of others, of which I may have to speak in connection with impotence as a result of continence, confirm me in the opinion that sperm once formed will, if it is not expelled in any of the usual ways, at first hinder, and finally stop, the production of new semen. It would indeed be interesting to know what becomes of this stored-up semen. The only chance for clearing up this question would seem to be an autopsy of persons who, after long continence, have died suddenly; but such an opportunity will seldom be offered. The third observation of these three individuals seems to prove that, in some cases of continence at least, the production of semen is not only not retarded but actually increased, so that the accumulation of sperm becomes so great that it causes the testicles to swell and become painful.

What is commonly called semen consists of the spermatozoids, formed in the testicles, to which are added the secretions of several glands situated in the terminal portion of the vasa deferentia, the so-called ampulla; there is added also the secretion of the walls of the vesiculæ seminales. The latter are not real glands but canals; their inner surface is much increased by folds and villi, and their secretion helps mostly in diluting the sperm.

It does not seem to me satisfactorily determined

PHYSIOLOGY OF THE SEXUAL ACT.

whether the *vesiculæ seminales* are secreting glands or merely receptacles for the sperm. The new experimental researches about the physiology of the seminal vesicles by Rehfisch [1] are very interesting, but do not reveal any important or new points. It is probable that the vesiculæ seminales and the ampullæ of the vasa deferentia have two functions,—viz., to serve as receptacles, and to produce an albuminous secretion that attenuates the sperm. The contents of the ampullæ and of the vesiculæ seminales of fresh cadavers have all the characteristics of ejaculated sperm, but are poorer in spermatozoids. Henle [2] endeavors to explain this fact by supposing that at the last moment of coitus that portion of the vas deferens nearest to the ampulla evacuates its contents more quickly.

To the semen, attenuated as above described, is later added the secretion of the prostata and of Cowper's glands.

Eckhard obtained by irritation, from the prostata of a dog, a secretion which had the specific gravity of 1.012, one per cent. of albumen and 2.4 per cent. of solid constituents.

The relation of the *prostata* with its secretion to the semen is not clearly demonstrated, as it is still a question whether the secretion of the prostata is added at all to the seminal fluid. Henle [3] is quite right when he says that the seminal fluid in the ampullæ of the vasa deferentia and in the vesiculæ seminales, although closely resembling the ejaculated semen, is poorer in sperma-

---

[1] Deutsche med. Wochenschrift, 1896, No. 16.
[2] Op. cit., p. 389.
[3] Op. cit., p. 401.

tozoids. Hence what needs explanation is not an attenuation of the semen, but an actual increase in spermatozoids. Henle says further: " It is scarcely reasonable to suppose that the function of the prostata is to attenuate the semen, as the principal mouths of the prostatic ducts are situated behind the summit of the colliculus seminalis, and the latter seems to shut off the urethra during erection. Finally, as far as my knowledge extends, the concentric concretions, which are seldom wanting in the glandula prostatica of old men, and which are often found in the outlets of the prostata, are not found in the ejaculated semen."

Henle asks, " Is the prostata connected perchance with erection, and does it furnish the mucous fluid which oozes from the mouth of the urethra after continued erection ?"

Finally, it is held that the mucus of *Cowper's glands* unites with the sperm, but Henle[1] thinks that Cowper's glands do not add their secretion to the sperm, and are, therefore, not to be reckoned among the organs of the sexual apparatus, but of the urinary apparatus. Beyond doubt these glands, together with the mucous glands proper of the urethra, make provision for the moisture and lubricity of the urethra, and thus, in any case, aid in ejaculation, and perhaps, as Hensen[2] conjectures, clear the urethra of remains of urine, which have an acid reaction and kill the spermatozoa. Besides this, we must not forget that Cowper's glands are perfectly developed in infancy.

Let us direct our attention now to the ejaculated

---

[1] Op. cit., p. 413.
[2] Op. cit., p. 101.

PHYSIOLOGY OF THE SEXUAL ACT.

sperm. The quantity and quality vary greatly in different individuals, as well as in the same individual, the differences being determined by the variable productiveness of the glands, which differ in the same manner and degree as their products, for reasons, some known and some unknown.

Mantegazza[1] finds the quantity of ejaculated semen of a thirty-year-old man, weighing about eighty kilograms, to be between 0.75 and 6.0 cubic centimeters. Ultzmann[2] estimates the average quantity for a moderately living man at ten to fifteen grams.

It is to be regretted that, in connection with the above figures, there are no indications how these ejaculations have been brought about, because those from pollutions are considerably less in quantity than those in coitus. This fact, which has never yet been noted, may be explained in this way: the vesiculæ seminales and spermatic ducts do not always empty their contents into the urethra at the same time, and it may be that when the excitation is moderate only one of the receptacles empties itself.[3]

The measuring of the quantity ejaculated during coitus is an impossibility, since the quantity is less when the coitus is prematurely stopped or interrupted, as in the cases known as "frauding," or "coitus interruptus." This may explain why the so-called "fraudeurs" can at first accomplish coitus oftener than those who follow nature, and why the same individual when he begins

---

[1] Ricerche sullo sperma umano. Gazzetta medica Italiana Lombardia, 1886, Nr. 34.

[2] Ueber Potentia generandi und Potentia coeundi. Wiener Klinik, 1885, Heft 1, p. 2.

[3] Landois, op. cit., p. 1024.

frauding can repeat the act more frequently than when he does not cheat nature in this way. The bad effects of frauding, also, which Bergeret[1] paints in colors decidedly too gloomy, may here find their explanation. There are cases, nevertheless, where long-continued frauding has undoubtedly weakened the sexual sense. I have observed this repeatedly within the last few years. The sexual sense soon recovers when frauding is discontinued.

The ejaculated sperm is a colorless, opalescent fluid, in appearance resembling boiled starch. The seminal filaments on their way from the testicles to the meatus urinarius externus are joined by the products of sundry glands, and thereby assume certain characteristics, the odor of chestnut-blossom or newly sawn bones being peculiar to the main substance formed in the testicles. The alkaline reaction comes, possibly, from the secretion of the prostata; the color, from the admixture of the secretion of the vesiculæ seminales; the gluey consistency, from the secretion of Cowper's glands.

The seminal liquid is heavier than water, soluble in water and acids, and coagulable by alcohol. Colorless, brittle crystals separate from it when it has been standing for a long time, the process being somewhat more rapid when the sperm has been placed in ice, or when inspissated. Vauquelin's chemical analysis is as follows:

Water	90
Organic substance	6
Earthy phosphates	3
Sodium chlorid	1

---

[1] Des fraudes dans l'accomplissement des fonctions generatrices. Paris, 1884.

PHYSIOLOGY OF THE SEXUAL ACT. 71

The consistency of the sperm changes soon after ejaculation, becoming more nearly liquid.

Whether or not a fluid is sperm can be proven only by microscopic examination. Sperm under the microscope shows seminal corpuscles, granules, cells, and epithelia from the prostata (?) and urethra. If the sperm comes from coitus, it also contains pavement epithelium from the female genitals.

Frequent indulgence will both diminish the quantity and impair the quality of the sperma. Differences have been observed in this with respect to individuals and to time, and I can myself assert as a result of numerous observations that habit and sexual vigor are the main agents producing such differences.

The presence of spermatozoa determines the fertilizing power of the semen. The sperma of a vigorous man in a state of perfect virility shows under the microscope a very animated life picture, which Ultzmann very appropriately compares to the life exhibited by an ant-hill. The spermatic filaments are seen to shoot across the field of vision like arrows, making movements that seem anything but aimless, so that it is not to be wondered at that they were at first thought to be animalcules.

This animation is most lively in newly ejaculated semen, and grows calmer in proportion as the spermatozoa die; but, if the sperma be preserved under favorable circumstances, some of the filaments may be seen alive even at the end of forty-eight hours. They are killed by urine and the vaginal mucus, which has an acid reaction. Mantegazza[1] found that the spermatozoa retain their vitality in temperatures varying from 15° to

---

[1] Loc. cit.

47° C. Heating above 47° or freezing below 15° kills them. It was further discovered that these filaments remain vital in frozen semen as long as six days, possibly even longer. Slightly alkaline fluids, such as blood, favor the life of the spermatozoids. Such fluids, as well as concentrated solutions of salt, sugar, and albumen, are capable of reviving spermatozoids that are already motionless. Water destroys motion after an hour, at the latest; but they are very lively in purulent sperm; consequently they are not affected by pus.

The dead spermatozoids show various figures of the lifeless tail. In those that die after ejaculation it is stretched out or slightly bent; whilst in those that are ejaculated dead it is spirally wound up, but seldom broken.[1]

Most authors adduce various experiments, the results of which have led to an opinion not wholly correct,— viz., that frequent coition reduces the number of spermatozoids, and daily coition makes them disappear altogether. To these investigators I can oppose many observations, where, after coition repeated daily several times, the sperma still contained numerous spermatozoids. These observations were made in the years 1884–1888. The differences in the results of the *observations* can easily be explained, inasmuch as examinations made on old men lead to different conclusions from those made on vigorous and perfectly virile young men.

Before I refer to some of my own observations, I wish to observe that of necessity it has not always been possible to make the microscopic examinations immediately after coition, nor at an equal length of time after copula-

---

[1] Compare Ultzmann, op. cit., p. 4.

tion; and, furthermore, the frequency of coition could not be regulated by the investigator. The sperma was invariably taken in the largest quantities possible, and was preserved in glass tubes, which were sealed and so placed that neither air nor light could affect the contents. All the microscopic examinations were made with a No. 3 ocular and a No. 7 objective of a Wetzlar microscope.

OBSERVATIONS ON SUBJECT NUMBER ONE.

A vigorous and perfectly healthy man, twenty-nine years old, who performed coition on an average once a day. The microscopic examinations gave the following results:

From October 1 to October 9 one coition per day; on October 9, examination of sperma nine hours after coition; result, few spermatozoids, mostly dead, not well developed; but the few still moving are very lively, although mostly young forms.

October 10.—One coition; sperma not examined.

October 11.—Sperma examined eleven hours after coition; result, few spermatozoids, all dead.

October 12.—Examination nine hours after coition; no spermatozoids.

October 14.—Coition after a lapse of forty-nine hours; examination eight hours afterward; result, about thirty spermatozoids, few alive, movement slow but very energetic. The coitus having been performed with a woman whose menses had begun, there is a blood-corpuscle (I could never discover blood-corpuscles in the ejaculations of so-called continent persons, as reported by Richard[1]); a few seminal filaments join the blood-cor-

---

[1] Histoire de la génération. Paris, 1883, p. 159.

puscle, but, after a momentary effort, push the corpuscle aside.

October 17.—Coitus after forty-eight hours' rest; examination eight hours later; result, about fifty dead filaments.

October 18.—Coitus after sixteen hours' rest; examination sixteen hours later; result, many spermatozoids had died, a few had been ejaculated dead.

On the same day, the sperm of a coitus, being the third within twenty-four hours, was examined ten hours after coition; result, very numerous spermatozoids, movement very energetic and as if with a purpose in view.

October 20.—One coition after an interval of forty-eight hours; examination ten hours after; result, fifty spermatozoids, all dead but a few. Second coition, one hour after first; examination nine hours later; result, spermatozoids very numerous and exceedingly lively; scarcely one-third had died.

October 21.—Coition after an interval of twenty-three hours; examination eight hours later; result, numerous spermatozoids, all living.

October 22.—Coition after sixteen hours' rest; examination sixteen hours later; result, fewer spermatozoids ejaculated alive, and, with few exceptions, all had died prior to my examination.

October 23.—Coition after thirty hours' rest; examination seventeen hours later; result, very many spermatozoids, but nearly all had died; besides a few ejaculated dead.

Similar results from examinations of sperm from same individual on October 24 and 25.

October 28.—Coition after sixty-four hours' interval;

examination eight hours later; result, about sixty spermatozoids, which had nearly all died, but one-third of them had been ejaculated dead.

October 29.—Coition after six hours' rest; examination two hours later; result, very numerous, active spermatozoids, only a few ejaculated dead; very few had died.

October 31.—Coition after fifty-eight hours' interval; examination sixteen hours later; result, spermatozoids few in number, nearly all ejaculated dead; the rest had died.

November 1.—Coition after fifteen hours' rest; examination one hour later; result, rather more spermatozoids, nearly all alive, though not moving with energy; others ejaculated dead.

On same day, a second coition after fifteen hours; examination ten hours later; result, spermatozoids numerous, lively, only a few ejaculated dead.

SECOND SUBJECT FOR OBSERVATION.

Powerful man, thirty years old, healthy, though inclined to obesity; enjoying life, very vigorous in sexual matters, and observes no rule at all; offers himself for ten days' observation.

February 10.—Coition after forty-one hours' interval; examination ten hours later; result, spermatozoids few in number, but moving with energy; few ejaculated dead, few died after ejaculation.

Same day, second coition, one hour later; examination nine hours after; result, spermatozoids very numerous, lively, and moving energetically; without exception ejaculated alive, and only very few had died.

February 11.—Coition after twenty-three hours' rest; examination ten hours later; result, spermatozoids less numerous; some moving, some had died, the rest ejaculated dead.

February 13.—Coition after an interval of forty-three hours; examination fifteen hours later; result, spermatozoids very few and most of them ejaculated dead; none moving.

February 14.—Coition after thirty hours' rest; examination and result, same as February 13.

February 15.—Coition after seventeen hours' rest; examination sixteen hours later; result, spermatozoids numerous, moving energetically.

Same day, second coition six hours later; examination ten hours afterward; result, spermatozoids not very numerous, but movement lively; most of the forms young; none were ejaculated dead.

February 17.—Three coitions within six hours; spermatozoids of the third coition examined ten hours after; very numerous, some of them still moving with energy; many had died, only few were ejaculated dead.

February 18.—Coition after eleven hours, therefore fourth coitus within seventeen hours; examination one hour after; result, spermatozoids less numerous, but moving with great activity and energy; few ejaculated dead, almost none died after ejaculation.

February 19.—Coition after fifteen hours; examination ten hours later; result, spermatozoids extraordinarily numerous, very well developed, with energetic and lively motion; none ejaculated dead, and only isolated ones had died.

THIRD SUBJECT FOR OBSERVATION.

Age not quite thirty, healthy and vigorous, life very active, but finds time to enjoy it, and makes best use of his leisure hours for venery.

After fifteen days' abstinence, intercourse fourteen times within six days. The last ejaculation examined scarcely one-quarter hour after; result, spermatozoids very numerous, many well developed, moving with energy and vivacity. The field of vision shows the picture that Ultzmann compared to an ant-hill.

From these observations and many more I have made, it may be concluded that, with persons who have accustomed themselves to frequent intercourse and have the power to do so, the number of spermatozoids increases with the frequency of coition, instead of decreasing, as supposed by the older authors. It may be concluded also that the spermatozoids become very numerous, well developed, lively, and energetic only when coition is repeated.

It is difficult to say to what this phenomenon is due, though it is probable that the sperm of the vesiculæ seminales, which is poorer in spermatozoids, is evacuated first, and that only after that, by the repetition of coition, come the contents of the vasa deferentia, and last of all those of the testicles. This circumstance may be of special importance in the consideration of certain cases of sterility.

Furthermore, such results justify the supposition that the spermatozoids which enter the vesiculæ seminales for storage lose their vitality in that canal gradually. According to the results of my observations, however, they lose it rather quickly. I am still further confirmed in this belief by the fact that, although I have very often had the chance to examine the semen from nocturnal emissions, sometimes within an hour after such emission, I have seldom found living spermatozoids in such sperma. Most of the spermatozoids, which were often found in great numbers, looked as if they had been ejaculated dead, while only few had the appearance of having been ejaculated alive and of having died on account of their low vitality. I must not omit the observation that the first portion of a pollution scarcely ever contained a living spermatozoid, whilst the second was more likely to show living forms.

In my numerous microscopic examinations of sperm I have made a few more discoveries, which I shall here adduce briefly. The energy of movement of a spermatozoon is most easily determined by gentle pressure on the cover-glass, which causes a current in the small quantity of seminal fluid between the two glass disks. If there are any spermatozoa moving with energy, they will swim unaffected by the current, and continue in their original direction, some swimming against the current.

The forward movement of the spermatozoids seems to me to be produced not by a whip-like motion, as has been stated, but rather by a regular, rudder-like action on the part of the tail. This action is seen plainly when a dying spermatozoon is watched; the movements grow slower and slower until they gradually cease altogether.

It is very interesting to watch how spermatozoa often meet with premature death. A spermatozoon swimming along quite energetically suddenly strikes some obstacle; the tail finds itself caught between masses of detritus, fragments of cells; the spermatozoon makes desperate efforts, moves spasmodically, and seems thus to use up all its vital power in a short time, as it quickly dies. Sometimes the spermatozoon succeeds in disengaging itself, but usually it suffers some injury. I once saw one with a sharp bend in the tail close behind the head swim on in a lively manner.

CHAPTER IV.

ETIOLOGY OF IMPOTENCE.

Among the circumstances that determine sexual vigor the foremost of all is the structure of the genitals. We shall not here take into consideration abnormal formations, as they will be treated specially, but shall limit our attention to genitals anatomically normal.

The *appearance* of the genitals varies considerably as regards size, form, and color. The differences are noticeable even in childhood, in which case they cannot be ascribed to extraneous causes. Small genitals are always a sign of insignificant sexual power, while large ones, remaining in proportional size during erection, indicate great power. There are genitals which during sexual repose show large dimensions, but which are flabby and appear large only in consequence of the extent of the vascular meshes of the cavernous tissue, and do not grow larger proportionally by the filling with blood during erection. In such genitals erections do not occur readily, and they accordingly indicate anything but sexual vigor. Such qualities belong to genitals which owe their size to sexual excesses committed before puberty.

Together with the necessary dimensions the texture of the penis must possess firmness. The testicle must be sufficiently large, firm, and insensible to slight pressure.

Besides this, the vascularity of the genitals is of importance. A pinky, transparent cutis of the glans and warmth of the penis indicate that a sufficient quantity of blood is present. A pale glans and a penis that feels cool indicate local poverty of blood as well as sexual weakness.

Finally, the *excitability of the nerve-ends* that spread in the glans is of consequence with regard to the qualities of the genitals. Where the glans is entirely covered by the prepuce its surface is very sensitive. Such subjects answer quickly to external excitation, but are seldom noted for their power, as coition lasts too short a time, and they are inclined to what is called irritable weakness. If the glans is covered by the prepuce only partially or not at all, its epidermis grows harder and less sensitive; external excitation affects tardily; a greater amount of irritation is required to complete the coitus, and the sexual organs themselves become more inclined to insensibility, which becomes more noticeable in riper age, when the central excitations grow fewer by degrees. Innate as well as acquired advantages or defects in the formation of the sexual organs influence their capability of action advantageously or otherwise.

Stronger pigmentation in the sexual organs is generally accompanied by greater capacity in venery, which fact is seen in negroes, who are, as a rule, endowed with large genitals. The well-known rule that brown-haired men have usually more sexual power than light-haired ones is admissible only in the comparison of men of the same race.

The *bodily structure* of a man is of the utmost weight with respect to his sexual capacity. It is self-evident that an individual who is in every respect healthy and vigorous will accomplish more in sexualibus than one sickly and weakly. Apparent anomalies are not wanting, and, indeed, we see often enough that decrepit and cachectic individuals commit considerable excesses in venery; but in the long run the weakling and the sick man will suffer more or less disaster, and only an energetic metabolism can for any length of time enable one with any effect to resist the manifold ravages of the waste of semen.

Due weight must here be allowed also for *hereditary predisposition*. There are families in which all the male members are distinguished for great sexual power; whereas the contrary may be noticed often enough in other families. I can only agree with Morrow[1] when he says. "The important rôle which heredity plays in determining disorders of the genital function has not been fully recognized nor sufficiently emphasized by writers upon the subject." Indeed, I expressed the same views years ago.[2] Idiosyncrasy, in fact, all the psychical qualities of a man, powerfully influence his sexual capacity.

Of great importance also is *the age*. Some authors allow a greater, others a lesser, latitude. There are severe moralists who will not allow sexual enjoyment before a man has completed his twenty-fifth year, and who say that he must be moderate even then, and desist from it when he is fifty years old.

The vast difference in the opinions held upon this subject evidently proves that no fixed rule can be laid down. Most authors have committed the error of allowing their personal experience to act as a criterion.

---

[1] Functional Disorders of the Male Sexual Organs. A System of Genito-Urinary Diseases, Syphilology, and Dermatology. New York, 1893, vol. i. p. 1001.

[2] Pathologie und Therapie der Männlichen Impotenz. Wien und Leipzig, 1889, pp. 45, 73–83.

The following general principles may, however, be stated:

Nature alone indicates the time when a man should satisfy sexual desire. The course of nature should in no wise be interfered with or anticipated. When nature has done her work of bringing the man to maturity, when the testicles have produced sperm, and that sperm is thrown out by pollutions, when the youth's whole being is undergoing a radical change, then I cannot understand why he should not satisfy that impetuous, irresistible longing; why he should be condemned, in the best part of his life, to become an onanist or to lose his power by pollutions.

Virility generally begins when the man reaches the age of eighteen years, increasing until he has reached his fortieth year. From that time it begins slowly but steadily to decrease, until in his sixty-fifth year it is usually extinguished. There are some who in their sixteenth year, and even before, are perfectly fit for coition, and a great many who preserve their sexual power to an advanced age; whereas, on the other hand, some hardly enter into puberty at the age of twenty-four, and are overcome by senile impotence before they are fifty years old.

Besides these congenital qualities, over which the individual has no control, there are many circumstances generally beyond his control also, but which he may, nevertheless, endeavor to bring about so as to influence his virility more or less.

Here we must, first of all, mention acquired deformities and diseases of the genitals or other organs; after that the manner in which one husbands the gifts of nature. Too much indulgence may be just as injurious as too little. Furthermore, there must be taken into account the influence of nutrition, of certain alimentary or medicinal substances, and of occupation and habits. Finally, there are many other things that have more or less influence on the strength of a man in sexual matters, and their connection with virility seems peculiar and strange because we cannot understand it.

The varying *influence of seasons*, for instance, modifies the sexual power, though this cannot be acknowledged unconditionally. Everybody knows that man is given to love in spring more than in any other season. Some French investigators (such as Villermé, Quetelet, and Smits) have worked out tables in which the order in which the months are named is indicative of the number of conceptions that occurred therein; but they cannot be accepted as valid, because such proofs, based on statistical calculations, are not always independent of chance. Nor can we accept as an absolute proof the fact that most crimes against morality are committed in the spring; as here, too, many other social conditions play a part not unimportant.

Besides the influence of season, we must bear in mind the momentary disposition and the varying feelings of inclination and disinclination, to which we cannot deny a rather strong control over virility.

After this brief enumeration of the most important circumstances capable of influencing sexual virility, we may pass on to a description of the individual forms of impotence.

A classification of the manifold forms of this disease, which has hitherto received so little attention, offers, for the present, insurmountable difficulties.

Krafft-Ebing's schema[1] of all the sexual neuroses is very ingenious. He distinguishes three kinds. They are—

I. Peripheral neuroses.
    1. Sensory.
        a. Anesthesia.
        b. Hyperesthesia.
        c. Neuralgia.
    2. Secretory.
        a. Aspermia.
        b. Polyspermia.
    3. Motor.
        a. Pollutions (spasm).
        b. Spermatorrhea (paralysis).

II. Spinal neuroses.
    1. Affections of the erection center.
        a. Irritation.
        b. Paralysis.
        c. Inhibition.
        d. Irritable weakness.
    2. Affections of the ejaculation center.
        a. Abnormally easy ejaculation.
        b. Abnormally difficult ejaculation.

III. Cerebral neuroses.
    1. Paradoxia.
    2. Anesthesia.
    3. Hyperesthesia.
    4. Paresthesia.

Ingenious as this classification is, it is of no practical use for our purpose.

---

[1] Psychopathia sexualis. Stuttgard, 1890, p. 24.

ETIOLOGY OF IMPOTENCE.

Furthermore, Eulenburg[1] considers that differences between "peripheral neuroses" and "spinoferic neuroses" of the genitals can hardly be determined except on paper. This is certainly neatly expressed, and holds good also with reference to Eulenburg's division of the sexual neuropathic phenomena, or with reference to any other possible or impossible division: they all look well on paper, but, in reality, chaos reigns.

There are, as we have seen, many causes that may lead to impotence; and although it is always reducible to partial or complete failure of erection, yet the accessory circumstances accompanying this main moving force are often various, according to the exciting cause. In consequence of this, the disease may present to a careful investigator very different aspects, and therefore will demand also very different forms of treatment, determinable by the actual causes.

The usual division of impotentia cœundi into an organic form, a psychical form, a form depending on irritable weakness, and a paralytic form is surely not sufficient, because there are so many varieties that cannot be forced into such a frame.

Beard[2] distinguishes the following forms:

1. Slight deficiency, both of desire and capacity.
2. Deficiency of capacity with increase of desire.
3. Profound deficiency both of desire and capacity.
4. Erectile power increased abnormally, but no discharge of seminal fluid.

---

[1] Sexuelle Neuropathie. Leipzig, 1895, p. 44.
[2] Beard and Rockwell, Sexual Neurasthenia, fifth edition. New York, 1898, p. 124.

Mantegazza[1] distinguishes even as many as ten degrees or grades of sexual capacity, but avoids the exceedingly difficult task of a classification of the different forms of impotence.

---

[1] Igiene dell'amore. Milano, 1881, p. 112.

CHAPTER V.

FORMS OF IMPOTENCE.

CONGENITAL MALFORMATIONS AND DEFECTS OF THE SEXUAL ORGANS.

Congenital malformations of single organs of the human body are, fortunately, very scarce. While in the service of the Croatian government I had to examine the conscripts. Among six thousand young men there were only five who showed malformations of importance, and in three only were the genitals affected.

As all these six thousand men were, without exception, over twenty years of age, and as many deformed individuals do not attain that age, it would not be safe to infer an infrequency of malformations; but we might rather conclude from it the frequency of malformation of the genitals, because it does not shorten life so frequently as malformation located elsewhere.

Hypospadia and excessive smallness are the most frequent of the malformations connected with these parts; while entire absence of these organs is the most infrequent.

In malformations which prevent copulation altogether, the outer attributes of virility are absent. The whole appearance of these unfortunate beings resembles that of a woman, and appearance, voice, and behavior indicate that the formation of the genitals is not normal.

In the exceedingly rare cases of **absence of the penis**

there could be no possibility of copulation. Entire absence of both testicles is quite as rare, and has, perhaps, never been observed in adults.

Extreme smallness of the penis alone, or of the penis and testicles, occurs now and then, and is noticeable either at birth or later as an arrest of development. If the testicles are normally developed, and only the penis has remained very small, the desire and relative capacity for coition may be preserved unweakened, but the result will be a failure.

Such individuals are well aware of their defect, and are with difficulty induced to have intercourse with the opposite sex, particularly after they have had some bitter experience. Most of them seek to satisfy in some other way the sexual desire that they may feel. Few of these wretched beings are fortunate enough to meet with a woman who possesses the ability to suppress her sexual passions sufficiently to enable her to live contentedly with a husband so deficient in the sexual organs, although the capacity of self-denial is a peculiarity inborn with the entire sex. If both penis and testicles are diminutive, there is a poor prospect for the sexual desire, such subjects, as a rule, never holding intercourse with the other sex.

The opposite deformity—viz., **excessive development of the penis**—occurs also, and is usually indicative of great sexual power. It offers no obstacle to coition, provided a proper mate is found, and, generally, little difficulty is experienced in this line.

The negro has a very large penis, but it does not increase during erection in proportion to its size when flaccid. The Japanese, on the other hand, has a small penis, even in proportion to his small stature.

It is very rare to find a marked congenital flexion in the penis arising from a deformity in the corpora cavernosa. Curvature to a very considerable degree would render copulation impossible.

Of more frequent occurrence is a **defective development of the erectile tissue,** in which case the penis may be sufficiently large, but abnormally flabby. This congenital defect is, in my eyes, of great consequence, and I cannot understand how it is that, in spite of its frequency, it is ignored entirely by modern authors. Lallemand alone has carefully described this condition.

Impotence is said sometimes to depend also on great **narrowness of the orificium externum urethræ.**[1] The possibility of this may be admitted the more readily since we know that stricture of the urethra undoubtedly causes impotence. I have had occasion to observe a man fifty-five years old whose orificium externum was exceedingly narrow, and who, though virile, could never impregnate a woman. It is more than probable that this malformation, unimportant as it may seem, was the cause of the sterility; for the semen was quite normal, and an examination of his wife brought no explanation to light.

Some consideration is also due to the state of the prepuce and the frenulum. Complete **absence of prepuce** probably occurs very seldom. Even among the Orientals and Hebrews, who have continued to remove it for thousands of years, this artificially caused absence has never yet become hereditary. Even if the absence of the prepuce were an effect of inheritance, it could

---

[1] Maximilian v. Zeissl, Ueber die Impotenz des Mannes und ihre Behandlung. Wiener medicinische Blätter. Wien, 1885, Nr. 15.

never be prejudicial to the capacity for coition. Possibly, as Roubaud[1] asserts, the glans grows therefrom less sensitive, the act less agreeable, and consequently the carnal appetite less keen.

More frequently occurs a superfluity of prepuce, causing **phimosis**. Even phimosis in the highest degree cannot have a damaging effect on the capacity for copulation; it can only interfere with the natural course, and make surgical help desirable. I must, however, call special attention to the fact that congenital excessive length of the prepuce is generally accompanied by a defective development of the member itself, so that, in some measure, the prepuce is only too long for the abnormally small penis, and then we have to deal only with the diminutive size of the penis. Too small an aperture through the prepuce often causes sundry diseases, especially of the nerves. In recent times, observations of this nature grow in number. In cases of congenital phimosis it happens now and then that the prepuce adheres to the glans even after puberty has been reached, and this would be a positive hindrance to coition.

In California I had an opportunity to observe a case of firmly adherent prepuce occasioned by a surgical operation for phimosis, when not enough of the prepuce had been removed, and, besides, the margin of the wound had not been stitched.

The **frenulum** is sometimes too long, sometimes too short. In the latter case it is a hindrance to erection. I have been compelled more than once to sever a frenulum that was too short. In one case the glans was

---

[1] Traité de l'impuissance et de la sterilité. Paris, 1876, p. 100.

drawn downward, and in another the frenulum tore during coitus, causing great pain.

Not infrequently there is a natural inclination toward fissures on the surface of the glans and the inside of the prepuce, whereby a temporary difficulty in copulation may be occasioned. I have seen a young university professor who, after every excess in venery, were it ever so slight, suffered from deep and gaping fissures on the surface of the glans.

Hypospadia of a high degree—*i.e.*, congenital opening of the inferior wall of the urethra—and the much rarer **epispadia**—*i.e.*, congenital opening of the upper wall of the urethra—may cause partial or absolute impotence. Hypospadia of the lowest degree—viz., when the orifice of the urethra is situated in the furrow of the glans, at the root of the frenulum—never interferes with copulation, but diminishes the chances of fecundation. In cases of hypospadia of a high degree, when the urethra opens as far back as at the perineum, and the member itself is very small, flattened, and bent downward, both copulation and fecundation suffer, as a matter of course.

Quite similar are the cases of **epispadia**. According to the degree of deformity, the member is at the same time shortened, flattened, and turned upward like a hook, so that introduction into the vagina succeeds only partially or not at all, and the injection of semen is accomplished with great difficulty. Exceptionally, impregnation may result in cases of either infirmity, even when of a high degree.[1]

Sundry circumstances may disturb the process of

---

[1] Hofmann, Lehrbuch der gerichtlichen Medicin. Wien, 1881, II. Aufl., p. 68.

"descensus testiculorum," so that either one testicle remains in the abdominal cavity or else both. The former case is called **monorchidia**, and occurs oftener than the latter, which is called **cryptorchidia**. Neither of these conditions causes impotence if it is not accompanied by defective growth of the testicles. It is even asserted that such hidden testicles produce more sperm because they are in a warmer location.[1] Cryptorchidia is said to cause sterility, but not impotence.

These two assertions are contradictory. Godard, Hunter, and Curling have in cases of cryptorchidia found no spermatozoa in either testicles or vasa deferentia, nor in the vesiculæ seminales. Contrary to this, Taylor[2] knows four cases of cryptorchidia where there are children; Pelikan, one case; and Beigel[3] found sperma in the semen of one affected with cryptorchidia.

It is beyond doubt that monorchidia cannot be injurious to potentia cœundi or generandi, because it seems, indeed, to be hereditary. I myself know a civil official in Europe, a monorchis with large penis, who was very vigorous in sexualibus until an advanced age, and late in life begat two boys, the younger of whom is also a monorchis. Each son has likewise a very large penis, and resembles his father in vigor in venery.

Young people afflicted with such defects usually become very unhappy the moment they are aware of their defect, because they believe they will have to renounce sexual enjoyment. In many places the popular belief

---

[1] D. Campbell Black, Human Anatomy and Physiology, Part VII. p. 6.
[2] Hofmann, op. cit., pp. 59, 60.
[3] Virchow's Archiv, Bd. cviii. p. 144.

accredits such monorchids with possessing extraordinary power.

As rare curiosa we have to call attention to **hermaphrodites**, or individuals who, by vicious conformation of the genitals, are hindered partially or entirely from indulging in intercourse; also the much rarer cases of partial or entire want of single parts of the apparatus serving as excretory ducts for the semen, such as the vasa deferentia, ductus ejaculatorii, and vesiculæ seminales. The rarest cause of impotence may be congenital azoospermia.

ACQUIRED DEFECTS IN THE SEXUAL ORGANS.

We mean by this the permanent defects. The various diseases of the sexual organs will be treated later on.

The entire or partial **loss of penis or testicles** will be the first subject of discussion. In the Orient there are even nowadays people on whom, during infancy, a most atrocious act has been perpetrated,—that of cutting away the entire external apparatus of generation. Such perfect *eunuchs* bring higher prices than those who are deprived only of their testicles. In Russia we have the *Skopti*, who, in insane fanaticism, voluntarily submit to such a mutilation.

Of course, copulation is out of the question in a case of complete absence of the external genitals; it is likewise impossible in the absence of a penis. Neglected venereal diseases sometimes destroy a part of the virile member, and malignant neoplasms sometimes make its removal a necessity. There are, besides, unfortunate accidents and traumatic influences that may cause loss of the penis. This is really a very pitiable condition, because the sexual appetite is left, while the possibility of satisfying it is gone. If, however, the traumatic action has left part of the penis, there is no impotence as long as the stump remains erectile. I knew a wealthy tradesman who had lost nearly the whole of the glans in consequence of a simple ulcerous disease which he had unwisely kept secret and neglected. After the wound was healed he had regular intercourse with his wife, but always took a long time to accomplish ejaculation. His wife complained that the friction of the skin and the cicatricial tissue caused her pain, wherefore she had always to apply a sufficient quantity of grease before copulation. The cicatricial tissue was not firm; in the flaccid condition of the penis it was very yielding. I never saw the stump during erection.

The **absence of testicles** may be due to various causes. First of all, they may have been removed surgically,—an unjustifiable operation, still practised in the Orient, and in Rome considered "ad majorem dei gloriam;"[1] they may have required removal on account of disease or the growth of some tumor; or, finally, they may have become completely atrophied from some cause or other, as, for instance, through syphilis, epididymitis, or pressure by a large hydrocele, varicocele, scrotal hernia, etc.

If the testicles are lost before puberty, both sexual desire and capacity for sexual gratification are impossible; whilst both may be preserved for some time, possibly for a long time, if the testicles are lost after puberty. Such cases as are recorded where women have amused

---

[1] Mantegazza, Gli amori degli nomini, vol. i. p. 175, says, "Man castrated himself and emasculated others, driven to this infamous mutilation by the most opposite reasons,—the desire to triumph over human weakness and aspire to heaven, revenge, jealousy, luxury."

themselves with castrated men[1] refer to individuals who had been emasculated only a short time, or, at least, after puberty.

At the University of Vienna, a fellow-student of mine had an obstinate epididymitis caused by gonorrhea that brought on conditions in consequence of which one of the testicles had to be removed, whereupon the other testicle atrophied. This unfortunate young man practised copulation for some years after this, boasted of it, and quite ostentatiously courted the ladies. Gradually his power of copulating vanished, and after three years he withdrew from the society of women altogether, and grew peevish and reserved, until one day he disappeared and was never after heard of. This case has left a vivid impression on my memory, and illustrates quite characteristically the influence of the virile power on the whole being.

Absence of one testicle, from whatsoever cause, leaves virility unaffected if the other testicle continues its functions. The Hottentots are said to amputate the left testicle of their youths before entrance into matrimony.[2]

Hydrocele and **inguinal hernia**, if of a high degree, may encroach upon the integument of the penis, causing that organ to disappear from view completely, thus producing a mechanical impediment to copulation. Tumors, by reason of their size, form, and position, may unfit the member for introduction into the female genitals.

Injurious influences of a traumatic nature, more rarely

---

[1] " Sunt quas eunuchi imbelles ac mollia semper
 Oscula delectent et desperatio barbæ
 Et quod abortivo non est opus."—JUVENAL.

[2] Mantegazza, Gli amori degli nomini, vol. i. p. 175.

diseases—as, for instance, those following gonorrhea—may produce **changes in the corpora cavernosa,** such as local obliteration of the meshy passages, nodi, and wheals or callosities. Under such circumstances the meshes will not all fill equally in erection, some parts remaining quite soft, whereby the member takes on a more or less bent form, and will thus become unable to penetrate into the vagina. Curschmann[1] mentions a pertinent and interesting case, not isolated in literature : A robust railway official, twenty-six years old, awakened one morning with a violent erection. He endeavored to bend the penis downward, when suddenly it gave way, caused great pain, and sank down. There was profuse bleeding underneath the skin of the penis, so that it was black and blue and almost as large as a fist when Curschmann saw the patient. After recovery it was discovered that by the violence practised the right corpus cavernosum had been torn, and, in consequence, at every erection the penis was bent upward and to the right, so that copulation became mechanically impossible.

Small fissures in the corpora cavernosa may be caused by violent motion during coitus. Such a case I had under my care in the year 1887. A restaurant-keeper, forty-one years of age, wanted, after a slight excess in Baccho, to do homage to Venus also. Both he and his wife were in a somewhat exalted mood, and probably proceeded rather impetuously. The husband told me that the erection was of unusual vigor, and just before ejaculation he suddenly felt a sharp pain, so that he had to discontinue the act. The erection subsided at once,

---

[1] Impotenz, Band ix. 2; Ziemssen. Handbuch der speciellen Pathologie und Therapie, p. 530.

but the appearance of the member was such that he was compelled to get medical advice. When I saw the man, an hour after the occurrence, the penis was much swollen, black and blue all over, only an irregular streak on the right-hand side having its usual color. I ordered cold applications and occasional painting with iodin. After ten days the swelling had disappeared. I could feel a somewhat hardened spot on the left corpus cavernosum, but there was no further interference with erection and copulation.

Finally, there are to be mentioned the so-called *penis-bones*. They are of very rare occurrence, and may arise through the ossification of single parts of the albuginea of the corpora cavernosa. In case they seize upon larger parts, they may prevent dilatation, and thus annihilate erection and the power of copulation.

Persistent changes in the mucous membrane of the urethra, such as strictures, frequently, but not always, cause impotence.

CONSECUTIVE IMPOTENCE.

The performance of coition requires all the power of the individual and, above all, a normal state of the whole body. The most various diseases can affect sexual vigor and even destroy it.

Of least account is virility in **acute diseases** of a serious nature. Sexual desire is heightened during the prodromal stage of most of the acute diseases as long as the approaching fever, which may be quite high, is only announcing itself by an incomprehensible agitation and general uneasiness. Such patients are sometimes carried along violently to sexual excesses. I have often watched this condition. Thus, for instance, a man, twenty-six years of age, accomplished coition, contrary to his usual habit, three times in the night before scarlet fever declared itself, as he ascribed heaviness in his legs and a state of excitement to unsatisfied sexual desire.

During the illness, on the contrary, the sexual appetite is nil. During convalescence every other desire will make its appearance before this one, and its reawakening is with justice greeted as the sign of returning strength. The patient requires his forces for other purposes during attacks of acute diseases, and his temporary stagnation in sexual activity is a wise provision of nature; it would be absolutely wonderful if, unfortunately, other functions of the body, and principally that of digestion, were not impaired at the same time.

After a protracted, severe, and exhausting illness, during which the reproduction of spermatozoids diminishes or may be altogether arrested,[1] impotence sometimes lasts a long time. All the other consequences of the illness may be overcome before this sexual weakness, so that there may be a danger of its becoming permanent. *Diphtheritis*, which is sometimes followed by protracted paralysis and muscular atrophy, causes impotence now and then.[2]

The assertion is met with in old books, and has been copied into modern ones, that persons affected with **phthisis** are generally apt to commit sexual excesses (phthisicus salax). I most emphatically deny this. A

---

[1] Rosenthal, Ueber den Einfluss von Nervenkrankheiten auf Zeugung und Sterilität. Wiener Klinik, 1880, Heft 5, p. 165.

[2] Hofmann, Lehrbuch der gerichtlichen Medicin. Wien und Leipzig, 1881, p. 66.

FORMS OF IMPOTENCE.

phthisic person may have acquired the habit of frequent sexual intercourse in former times, and then, during his illness also, may go to excess for a time; but, surely, these are exceptions. As a rule, phthisics are not much inclined to physical love, and this is in keeping with the condition of their physical strength. For the purpose of noting this feature I have for many years past carefully watched and examined numerous phthisical patients. Without exception and without regard to age, they all entirely renounce sexual gratifications without experiencing the least difficulty. Indeed, quite young husbands affected with phthisis practise copulation very infrequently even during the intermissions of their illness.

Chronic diseases impair the sexual power proportionately to their effect on the rest of the body. Chronic diseases in the organs of respiration and of digestion affect virility only in the same measure as they debilitate the body in general and lower vitality. Disease of the heart does not impair virility until it amounts to a serious disturbance of the circulation. In the first stage of the disease the patient is rather nervous and inclined to sexual excess. The same is true with those suffering from disease of the liver.

Virility is affected also by some general diseases. Impotence is particularly frequent with persons suffering from **diabetes**, and often constitutes one of the first symptoms of the disease, noticed long before the physical being commences to deteriorate. The seminal secretion is also said to stop in this disease.

As a special enemy of the virile power must be mentioned **obesity**, which has only recently been studied carefully as a disease. In exceptional cases obese people may be very powerful in sexualibus. I myself have known such cases, but obese persons are usually fond of their comfort, in copulation as in everything else. They frequently prefer the pleasures of the table to those of love, and, moreover, are not much troubled by sexual desire. They give off but a scanty secretion of sperm, often suffer from adipose degeneration of the testicles, and are apt to become completely impotent. Kisch found invariably that the sperm of nine out of ten obese men showed under the microscope only molecular detritus and sperm-crystals, but no spermatozoa at all. In case of obesity of a high degree, especially when there is a *pendulous abdomen*, copulation may be mechanically impossible.

Anemia may be a cause of impotence, like any other disease carrying in its train debilitation of the body in general. If anemia is acute, it is generally accompanied by impotence, whilst anemia of a chronic nature causes at first only weakening of the sexual capacity. According to Roubaud,[1] *chlorosis* is also a cause of impotence, but the case he describes is such that I should certainly characterize it as neurasthenic.

It would seem almost incredible that a **severe cold** could affect virility seriously, if the fact were left out of consideration that the sense of smell influences the sexual appetite. In a man really vigorous sexually a cold, be it ever so severe, cannot annihilate virility, though it lessens the sexual desire. Schiff has removed the nervi olfactorii in new-born dogs, after which the male was unable to find the female. Mantegazza deprived rabbits of both eyes without any effect on copu-

---

[1] Op. cit., p. 213.

lation.[1] In man the olfactory sense has not so great a power as in animals, because a great share of influence over the center of erection has been allotted to other senses as well as to the developed faculty of thinking; nevertheless, the sense of smell is very important.

Impotence is very frequently a symptom of **disease of the** central or peripheral **nerve apparatus**. This is clear to any one even without a thorough knowledge of physiology. The practitioner sees typical examples almost every day. Nearly all diseases of the brain and of the spinal cord have great influence over the virile power,— some only for a time, others permanently, according to the character of the disease in question. Some of these diseases cause at first increased sexual excitement, which, in the further course of the disease, is followed by diminished sexual power or absolute impotence. A given disease of the brain and spinal cord does not always have the same effect in this respect. Apoplexy of the brain, for instance, may cause frequent erections at one time and entire loss of sexual desire and power at another.

In the first stage of *tabes dorsalis* the patient experiences for the most part an increased sexual desire, in consequence of irritation of the nerve-fibers which innervate the sexual apparatus; but later on sexual vigor gradually diminishes until it is entirely extinguished. Cases are known, nevertheless, where tabes existed in a high degree and yet the patient was still in some measure virile.

From this little we may conclude that there are no absolute rules to be given about the state of virility

---

[1] Igiene dell'amore. Milano, 1881, p. 277.

in affections of the brain and spinal cord. The physician has to examine every individual case and determine his treatment accordingly. In these diseases the medical man has to direct his attention to other things; he has no time to think of sexual virility, especially when life is in danger; and, besides, the patients—as the ataxic, for instance—care very little for the sexual power, for, as a rule, the desires are silent.

Impotence seems to me still more insignificant in some forms of *insanity*, excepting those of perverse sexual sensation, of which we shall speak hereafter.

Lesions of the brain or spinal cord may affect virility according to the spot injured. In the literature of the subject we read of cases where injury of the cerebellum has brought about a loss of sexual power. In certain injuries of the spinal cord, principally those of the inferior parts, and especially in spinal concussion, priapismus has been noticed. It constitutes one of the most troublesome symptoms, and defies all remedies. Lallemand[1] reports a very characteristic case. Rosenthal[2] mentions a case he has observed in which there was a fracture of the fourth, fifth, and sixth cervical vertebræ, with paralysis and anesthesia of the legs and trunk, together with retention of the urine and feces, and priapism for seven days during life and thirty-six hours after death.

Those cases in which impotence is to be considered as a symptom of some disease of the entire nervous apparatus have more importance for us, because they

---

[1] Des pertes seminales, tome ii. 1re partie, p. 64.

[2] Ueber den Einfluss von Nervenkrankheiten auf Zeugung und Sterilität. Wiener Klinik, 1880, Heft 5, p. 145.

FORMS OF IMPOTENCE. 103

occur more frequently, are of more consequence to the patient, and, finally, because therapeutic action is attended with more chance of successful result. In these cases no change in the nerve substance can be seen either macroscopically or microscopically. The pathological change consists simply in the altered capacity of nerve-action.

In the first place must be mentioned general nervousness, or, to call it by the name that originated in America, **neurasthenia**, or, as Rosenthal[1] calls it, depressive spinal irritation, all of which names are entirely appropriate. Neurasthenia has grown into a fashionable disease in this age of electricity, when every one belonging to the upper class has to hurry from early morning till late at night in order to accomplish his measure of work or of pleasure; when the everlasting hurrying begins in infancy and still continues during old age. By these psychical excitations, which exert such a frequent and lasting effect, the center of the vaso-constrictors particularly is kept in a state of irritation. Beard describes the conditions very forcibly when he says,—[2]

"The Indian squaw, sitting in front of her wigwam, keeps almost all of her force in reserve. The slow and easy drudgery of savage domestic life in the open air—unblessed and uncursed by the exhausting sentiment of love, without reading or writing or calculating, without past or future, and only a dull present—never calls for the full quota of her available force; the larger part is always resting on its arms. The sensitive white woman

---

[1] Op. cit., p. 142.
[2] Beard-Rockwell, Sexual Neurasthenia, fifth edition. New York, 1898, p. 59.

—pre-eminently the American woman, with small inherited endowment of force, living in-doors, torn and crossed by happy or unhappy love, subsisting on fiction, journals, receptions, waylaid at all hours by the cruellest of robbers, worry and ambition, that seize the last unit of her force—can never hold a powerful reserve, but must live, and does live, in a physical sense, from hand to mouth, giving out quite as fast as she takes in, much faster oftentimes, and needing long periods of rest before and after any important campaign, and yet living as long as her Indian sister—much longer, it may be—and bearing age far better, and carrying the affections and the feelings of youth into the decline of life."

Neurasthenia is either congenital or acquired. According to the rules of heredity, neurasthenic parents have neurasthenic children, or, rather, they have children naturally inclined to neurasthenia, which will tend to develop in them more and more under the least favorable circumstances, and who, if their regimen is not strictly regulated, will exceed their parents in neurasthenia.

General nervousness with its manifold symptoms is by no means a rare disease. The symptoms are frequently held to be special diseases, and it is only an apparent extreme if Beard and his disciples trace most of the diseases back to neurasthenia. It is an established fact that this neurasthenia is of much more frequent occurrence in America than in Europe. Explanation of this circumstance suggests itself when we compare the mode of living on the two sides of the Atlantic.

Beard[1] gives the following definition of neurasthenia: "Neurasthenia is a chronic, functional disease of the

---

[1] Beard-Rockwell, op. cit., p. 36.

nervous system, the basis of which is impoverishment of nervous force, deficiency of reserve, with liability to quick exhaustion, and a necessity for frequent supplies of force; hence the lack of inhibitory or controlling powers, physical and mental,—the feebleness and instability of nerve action and the excessive sensitiveness and irritability, local and general, and the vast variety of symptoms, direct and reflex."

According to Arndt,[1] neurasthenia is increased or decreased excitability and irritability in conjunction with incapacity to resist external influence,—*i.e.*, weakness in general.

The symptoms of neurasthenia are, according to Beard, the consequence of reflex irritations which pass not merely through the ordinary sensory and motory nerves, but also through the sympathetic system and the vasomotor nerves. The reflex irritation can start from any part of the body and pass over to another, but the brain and the digestive and reproductive systems are to be considered as the main seats. The symptoms of neurasthenia are inconstant and surprisingly interchangeable. Erb distinguishes between a " cerebral," a " spinal," and a " universal" neurasthenia; Beard recognizes, moreover, a " sexual" neurasthenia.

According to Beard, the sexual nervous exhaustion may be considered as cause, effect, or accessory to the other kinds of neurasthenia, but must, nevertheless, when fully developed, be distinguished from them just the same as general neurasthenia is to be distinguished from hysteria, hypochondria, and the various organic diseases of the nervous system with which it was con-

---

[1] Die Neurasthenie. Wien und Leipzig, 1885, p. 40.

fusingly mingled until quite recently. Beard considers sexual neurasthenia in general, and particularly in reference to its various complications, almost the most important of all the forms of neurasthenia. It must attract our notice that he further asserts that the clinically connected local conditions of sexual weakness in man, such as impotence, spermatorrhea, and the "irritable prostata," are to be looked upon merely as symptoms of sexual neurasthenia. This would, indeed, make matters easy, and the single word "neurasthenia" would explain many a thing that has appeared quite inexplicable until now. We might, however, at best, consider as symptoms of an existing sexual neurasthenia only those cases of impotence and spermatorrhea in which the organic conditions and their pathological alterations offer no hold at all for an explanation of the disease.

Beard states that the causes of sexual neurasthenia are: unfavorable social conditions, sexual excesses, immoderate use of alcohol and tobacco, special irritants, grief, and even climate; but, above all, he believes that the most prominent and predisposing causal force is modern civilization in regard to its wants and claims that are increasing from day to day.

The attentive reader of the excellent work of Lallemand "On Spermatorrhœa" will not fail to notice that it contains a similar idea, and I am inclined to agree with Lallemand rather than with Beard. Lallemand quotes observations where nearly incurable cases of spermatorrhea or impotence have followed insignificant causes—as, for instance, slight sexual excesses—which would have had in other individuals either no effect at all or, at least, an effect of no consequence. Lallemand explains this by assuming a natural nervous predisposition to what is now called neurasthenia.

Beard asserts that neurasthenic individuals are able to accomplish fatiguing mental work for years, and often during the whole period of life; so that, sometimes, nervousness and neurasthenia are associated with an enormous capacity for mental exertion. To illustrate this assertion, Beard says that it was by neurasthenic authors that the epoch-making works were produced; and names like George Eliot, Darwin, Heine, Spencer, Edwards, Kant, Bacon, Montaigne, Joubert, Rousseau, Schiller, and many more of the same rank illustrate beautifully the truth of the sentence that it is possible to produce works of genius and of consequence even with a limited quantity of nerve-substance and nerve-power, attended by a rapid consumption of the same. It seems to me fair, however, to question whether all the above-named celebrities were really neurasthenic, and also whether their neurasthenia was not more likely to have been caused by their colossal mental efforts than to have produced such works.

Beard's method of reducing to neurasthenia nearly all the pathological states in the system of reproduction would at once rid us of a great number of difficulties that assail every one who studies impotence. It would do away with the difficulty of arranging the different forms of impotency according to some logical system; because we should have only to differentiate a neurasthenic from an organic impotence, and, moreover, the diagnosis would also be an easier task; but, alas! the facts oppose such a simplification, and we have nothing to do but to continue the old patchwork now in use as well as we can. We can agree to the consideration of neurasthenic sexual weakness as a phase of a spinal or general neurasthenia, and must look upon an independent or self-subsisting neurasthenia sexualis as a form of impotence occurring frequently enough.

Very often virility is affected by **diseases of the sexual organs**; but let it be stated from the very beginning that this influence is in most cases a secondary one. Especially diseases of the colliculus seminalis and of the canals and apertures discharging into it cause, as a first effect, irregular involuntary losses of semen, and in this manner, indirectly, slowly, but surely, exhaust the sexual power.

Several diseases of the sexual organs have a direct influence on virility. Thus, *wounds and ulcers* of the penis are a direct obstacle to the accomplishment of coitus, whether they are of a specific nature or not. The same may be said of the condylomatous proliferations of a high degree, because every erection, and still more, friction against the female pudendum, causes great pain.

Some individuals suffer much from nearly continuous formations of herpes on the prepuce, the vesicles of which afterward turn into little sores. Uncleanliness is not always the cause of this disease; it is sometimes the consequence of an ulcus molle. Some people suffer from it without any apparent cause after every sexual connection, such vesicles forming on the member and preventing coition for a time. The very rare preputial calculus ("calculs du prépuce," Roubaud) may also form an obstacle to coition, but this is removable, and it certainly occurs only in consequence of great uncleanliness in connection with phimosis. Let me mention here that there are people who, seized by violent sensual desires, are capable of bearing even great pain; they feel no scruple in satisfying such imperious instincts, although thereby they harm themselves as well as others by increasing their own suffering and propagating infectious diseases.

Gonorrhea also is in reality an obstacle to copulation; but it is an obstacle about which people of the above description care very little. The disregard of this obstacle is of so much greater frequency as, unfortunately, the sexual impulse is so much stronger during blennorrhea, particularly in its acute stage. This circumstance contributes greatly to the dissemination of the disease. In the chronic forms of gonorrhea the state of irritability of the urethra, and therefore of the whole sexual apparatus, exists only slightly or not at all, but is sometimes roused to unusually violent manifestations by the locally applied remedies, such as caustic injections, etc.

Other obstacles set up by gonorrhea to the accomplishment of copulation are the further complications and the higher degrees of development which the disease often produces, and the extension of the blennorrheic process, such as inflammation of the prostata, the vas deferens, the epididymis, and, more rarely, the vesicula seminalis. Any one afflicted with such a disease will, however, not readily yield to the temptation of coition.

Virility is affected differently by the many diseases of the *prostata*,[1]—viz., acute and chronic inflammation, prostatorrhea, tuberculosis, abscesses and ulcers, hypertrophy and atrophy, cancer, tubercles, cysts, concretions, calculi, etc.

The question has been asked, Is the prostata after all

---

[1] Thompson, Diseases of the Prostata. German edition. Erlangen, 1867, p. 40.

of special importance in generation? "I would draw attention to Mr. Ellis's important paper 'On the Muscular Arrangements of the Genito-Urinary Apparatus,' wherein is remarked, 'I would propose the name orbicularis vel sphincter urethrae for both the prostata and the prolongation around the membranous urethra. while I would confine the old term, prostate (without the word gland), to the thickened and more powerful part near the neck of the bladder.'"[1]

Of the diseases of the prostate only acute inflammation and atrophy cause temporary or permanent impotency, whilst chronic inflammation for the most part diminishes sexual inclination considerably. The other diseases of the prostata have not been studied much in their influence on virility, as they generally make their appearance at an age when virility is of but small consideration.

Diseases of the *urinary bladder* are apt to increase sexual desire temporarily, catarrh of the bladder, for instance, being frequently followed by greatly increased libido sexualis, which degenerates now and then into priapismus and satyriasis. This increase in sexual desire is still more frequently met with when the neck of the bladder alone is diseased, and in such cases the ejaculation is, as a rule, accompanied with pain. Urinary calculi also nearly always cause increased irritability in the sexual organs. Roubaud[2] says that the very rare prolapse of the urinal bladder through the inguinal canal causes impotence, principally through the retraction of the penis occasioned thereby.

---

[1] R. Harrison, On some Points in the Surgery of the Urinary Organs. The Medical Record, 1888, No. 3.
[2] Op. cit., p. 255.

FORMS OF IMPOTENCE. 111

Strictures of the urethra have different effects on virility. They very often cause impotence, and are, moreover, always a hindrance to fecundation.

Superexcitation from disease of the sexual organs is generally followed by relaxation. The patient who suffers from gonorrhea is sexually excited; he could perform coition oftener than in the state of health if he were not withheld by the great suffering and the fear of the consequences; but when this state of irritation is past, reaction will set in. The unusual continence to which vigorous young men are forced by gonorrhea neutralizes the injurious influence of over-excitement of the nerves, and, possibly, of the specific action of the virus of gonorrhea on the sexual nerves. Sometimes we see, as Ultzmann [1] says, that such patients who have formerly been virile in a high degree become temporarily impotent after an attack of gonorrhea, especially when the disease is accompanied by catarrh of the bladder, prostatitis, or orchitis. Ultzmann thinks that in such cases gonorrhea has had a paralyzing effect on the nervous apparatus of the prostata. This conjecture must be admitted as perfectly well founded, but it does not exclude the possibility that the sexual nerves have become temporarily neurasthenic in consequence of the strain of the almost continuous state of excitement.

The invention of the endoscope has helped us to a knowledge of a number of diseases of the mucous membrane of the urethra, among which the affections of the colliculus seminalis and of its adjoining parts interest us in the first place, as they exercise great influence on

---

[1] Potentia generandi und Potentia coeundi. Wien, 1885, p. 24.

virility. Grünfeld[1] tells us that endoscopic observation of the colliculus seminalis in different individuals discloses varieties differing widely in color, size, consistency, and vascularity. These differences rest on a pathological basis. Continued investigations have proved that the various diseases of the sexual functions of man are not to be reduced to affections of the nerve-apparatus exclusively, but that they may also be dependent on structural diseases of the colliculus seminalis. We observe hyperemia, catarrhal swelling, and hypertrophy of the colliculus seminalis. Hyperemia is almost constantly met in onanists, while patients afflicted with spermatorrhea and impotence labor under various grades of catarrh of the colliculus seminalis.

Other diseases of the sexual organs do not affect virility. Cancerous or tubercular degeneration of the testicles does not impair the sexual power. I have often had opportunities to examine the ejaculated semen of persons afflicted with tuberculosis of the testicle, and never found spermatozoa therein; but the patients, subject almost without exception to frequent pollutions, were perfectly virile. Krzywicki found that tuberculosis of the male sexual organs mostly begins in the prostata; less frequently affected are the seminal vesicles, the epididymis, and the vas deferens; seldom the testicles; and very rarely the penis and the urethra.[2] Klebs, on the contrary, claims[3] that the tubercle-bacillus develops

---

[1] Die Endoskopie der Harnröhre und Blase. Deutsche Chirurgie, Stuttgart, 1881, Lief. 51, p. 172.

[2] Dr. Hermann Dürck, Über den gegenwärtigen Stand der Tuberkulose-Forschung. Wiesbaden, 1897, p. 354.

[3] Ibidem. p. 355.

largely in the testicular tissue, on account of the richness of the blood and lymphatic vascularity.

Some old French authors, and among them Lallemand, assert that *varicocele* can cause impotence. Lallemand[1] says even that in many cases of varicocele he found the testicles very small and soft, and, in case the veins of one cord only were affected, the corresponding testicle would be less fully developed than the other. Daily experience is often in opposition to this assertion. In many cases of varicose affections of the veins of the spermatic cords I have found the testicles rather enlarged than otherwise. Virility I never found impaired, although patients of this kind are not exactly vigorous in sexual matters, and are very much predisposed to gonorrheal and traumatic inflammation of the epididymis.

There is a great number of **poisons, medicaments, and foods** which diminish virility temporarily or permanently. Some manifest their injurious effect after a short and moderate use; others do not show themselves until after a longer or immoderate use. In this respect we meet the most contradictory statements in medical works. Very often one author quotes the assertions of another. These contrasts are, however, easily explained if we take into account the difference of the effect of one and the same medicament on different individuals. Take, for instance, quinin, the remedy most generally used until recent times. The same dose given to two equally vigorous men may cause in the one scarcely a slight tinnitus aurium and in the other the most unpleasant effects.

Alcohol, especially, exhibits its action on different in-

---

[1] Op. cit., p. 171.

dividuals differently. There are people who in a rather high degree of intoxication can accomplish the act of coition, whilst with others the sexual organs are completely paralyzed by the consumption of so small a quantity of alcohol that it would not affect at all the functions of the other organs of the body. Thus we see that in this respect it would not be safe to lay down general rules and apply them in all cases; but we would rather make this the subject of our next discussion.

Alcohol in general diminishes the sexual power, according to the strength of the article consumed. Alcohol is least concentrated in beer, and yet beer is, as a matter of fact, very unfavorable for virility. Gallant ladies are well aware of this, and it is only an exception for them to serve beer to their lovers. Too hasty an ejaculation may be delayed by a moderate consumption of beer,[1] while an intemperate absorption of the same liquid will hinder erection.

But why is beer disadvantageous to coition? Alcohol is of consequence only when consumed in large quantities, and yet we see that very light qualities of beer are perhaps worse for virility for the time being than heavy beer. Lupulin cannot be of importance either, as the quantity is too insignificant, and in cases where the efficacy of lupulin is desirable, it is generally without effect. Therefore the effect of lupulin is not so prompt and sure as that of beer. I think that, in an immoderate consumption of beer, both alcohol and lupulin (?) are of less importance than the great quantity of liquid consumed, which causes frequent urination and has a relaxing effect on the parts under consideration. Dr. Lehmann,

---

[1] Curschmann, op. cit., p. 535.

of Munich,[1] ascribes the diuretic effect of beer to the greater quantity of liquid, assisted by the influence of alcohol upon the heart. It is noticeable that erection is slower immediately after urination than some hours later. This observation suggests that sexual organs active in erection may be temporarily disturbed in their function by the evacuation of large quantities of urine, repeated at short intervals.

A moderate consumption of beer is rather advantageous for the act of coition, and, in wedlock, where I ascribed the absence of children to a possible frigidity of the wife, I have advised the husband to take some beer before coition, because I thought that by thus prolonging the act the wife might be roused out of her reserve and become more liable to conception. The result has justified this supposition in one case at least. These observations would also help to explain the large families in countries where beer is the habitual beverage, as in Bohemia and Bavaria.

The effect of wine in this respect is exceedingly different, varying with the kinds of wine as well as with the individuals consuming them. Here again my observations have proved that wines with diuretic tendency affect the sexual capacity for the time being more than do other wines. Some strong, dark-colored wines, such as certain kinds of Californian, Bordeaux, Malaga, Dalmatian, Smyrnian, etc., and also some of the stronger white wines, especially the Muscatel, if consumed moderately, have almost an aphrodisiac effect; whilst others, and particularly champagne, exercise an almost paralyzing

---

[1] Die Ursache der bekannten diuretischen Wirkung des Bieres. Wiener med. Presse, 1887, No. 42.

influence on the centers of erection, or, directly, on the apparatus of erection, as they increase the libido sexualis, but check the erection.

In brandy and liquors the quantity of alcohol only comes into consideration, and, though it may be larger than that in wine or beer, it tends to increase rather than diminish the sexual power, if the same time for subsidence has been allowed in both cases of consumption.

At any rate, the ancient Latins were right only as far as the woman is concerned when they said, "Sine Cerere et Baccho friget Venus," "Luxuriosa res vinum," "ut vino calefacta Venus, tum sævior ardet luxuries," etc.

Considering the first exciting and subsequent sedative effect of **coffee** and of **tea,** an immoderate use of the same might injure virility.

Some authorities assert that **smoking** is injurious to virility, but it is very difficult to form a correct opinion. Schtscherbak's searching investigations regarding the influence of tobacco on the nervous centers[1] have resulted only in the assertion that immoderate smoking, like the internal use of nicotin, undoubtedly affects the nervous centers; but it is difficult to determine what influence is exerted on the centers of the sexual functions. In acute intoxication with nicotin, copulation is out of the question, but when the symptoms of poisoning have passed, virility is exactly in the same condition as before. Chronic intoxication with nicotin, to which one may be addicted for many years with impunity,

---

[1] K voprosu o vlijaniji nikotina i kurenija tabaku na nervnie centri. Vratch, 1887, Nos. 4–9.

seems not to injure virility, as very great smokers may indeed be quite as great in sexualibus. A thirty-year-old Servian told me he had by experience found that his virility was seriously injured when he discontinued smoking cigarettes. This might, however, be merely imaginary.

The habit of **snuff-taking**, now becoming less common, is more likely to injure virility, as it weakens the sense of smell, and the odor of woman plays an important rôle in sexual matters, as is well known. Galopin,[1] speaking of snuff takers, says, "If they are gourmands, they deprive themselves of the bouquet of their dishes and wines; if they are young and vigorous, they deprive themselves of the pleasant odor of a beloved wife or mistress, as well as of a thousand pleasures which the olfactory sense of a clean and healthy man procures." But snuff users have little chance with ladies nowadays, as they unquestionably diffuse a disagreeable odor about themselves.

There are **foods,** solid and liquid, that are said to cause temporary impotence, but, in my opinion, this rests more or less on popular belief only; and, after all, virility must be at a low ebb when it can be checked by eating Lima beans, lettuce, etc.

There are many popular means of subduing amorous desire for a time. In Bosnia the moon-flower, under the name of "Neven," is highly prized as a powerful anaphrodisiac. Women make their husbands take it in the form of medicine, and they also put the blossoms among the linen of husbands about to go on a journey.

In France **digitalis** is said to have similar renown, and

---

[1] Le parfum de la femme. Paris, 1886, p. 39.

Campbell Black finds this virtue of digitalis very comprehensible, as it stimulates Remak's fibers.

We meet with the most contrary statements about medicaments reputed to have an injurious influence on virility. Our best observations are on **morphin**, which, according to Levinstein,[1] after first increasing sexual excitability, affects it finally in the opposite manner. An injection of morphin always has, on persons who are not accustomed to it, the effect of increasing sexual excitement and vigor. Rosenthal[2] states that injections of morphin of medium strength (0.03–0.06 per day) produce unusual hilarity and affability, heightened sexual excitability, increased refinement of the sense of touch, etc.,—all symptoms little known and appreciated.

In Persia opium is said to be used as an aphrodisiac.[3] This is in accord with the observations made on opium-smokers, who are extraordinarily vigorous sexually at first, their virility beginning to fade when the general marasmus always following this fatal habit reaches a certain degree.

Morphin, opium, and cannabis Indica have long since ceased to be used for therapeutic purposes only, there being a large number of persons for whom the very extensive use of these drugs has become an indispensable necessity. Opium-smokers and hashish-eaters sing real hymns[4] of praise to these poisons. These poisons are, indeed, pleasant and sure means of suicide for unhappy people or people weary of life in this world.

---

[1] Morphiumsucht. Berlin, 1887, p. 93.

[2] Untersuchungen und Beobachtungen über Morphiumwirkung. Wiener med. Presse, 1886, No. 49.

[3] Rosenthal, op. cit., p. 147.

[4] "Oh, just, subtle, and mighty opium!"

Their habitual use can be recommended to those only who desire to commit suicide. Paralyzing the inhibitory nervous centers in the brain, they probably thus increase virility at first. Dr. L. Passover[1] has observed that the long-continued use of morphin leads to atrophy of the genitals.

There are also the most contradictory statements concerning the influence of **arsenic** and its preparations on the virile power. Although arsenic has a different effect on different individuals, and may therefore affect their sexual powers differently, yet I must state that in the very frequent use I have made of arsenic in various diseases, even for years continuously, I have never observed a diminution of sexual vigor. I never saw a change in the sexual power of men or in the amorous desire in women, even in cases where arsenic had some disturbing effect, or where it did not produce the desired change in the disease under treatment, and its use had, in consequence, to be discontinued. On the contrary, several of my patients, who owe the return of health to arsenic, have with their health also recovered their virility; so that I do not hesitate to recommend arsenic, in conjunction with other remedies, of course, in certain cases of prostration, and also in impotence when it has been brought on by such prostration.

Rosenthal[2] asserts that arsenic exercises an unfavorable influence on the sexual power after continued use; but this is so only with the inhabitants of towns, whilst the inhabitants of Alpine regions have children in spite of the consumption of considerable arsenic. This, in-

---

[1] Wiener med. Presse, 1893, No. 7.
[2] Op. cit., p. 151.

deed, seems too improbable. The same writer observed recovery of sexual sensibility at the beginning of the use of arsenic.

Lead-poisoning, especially if acute, occasionally causes impotence. As physician of a large association of typographers I had, for a number of years, the opportunity to convince myself of the excellence of the observations made by Tanquerel des Planches, and of those made more recently by Roubaud.[1] I never noticed a diminution of virility in chronic lead-poisoning that was not accompanied by some other effect on the nervous system.

Long-continued use of **iodin** can produce atrophy of the testicles, besides that of other glands; but such cases are exceedingly rare, and entirely denied by some very experienced syphilologists. I have recently had two cases where the use of potassium iodid had exerted a very unfavorable effect on virility.

Prolonged use of **mercury** also is said to lead to atrophy of the testicles. Roubaud[2] has observed this in the case of laborers who work with mercury.

Salicylic acid and its preparations unquestionably impair sexual vigor, but only temporarily. My observations have convinced me that men are temporarily more or less impotent during the use of salicyl and its salts, which are so frequently employed. The experiments made by Kolbe and Dr. Lehmann,[3] in Munich, to prove the harmlessness of salicylic acid, have not had any results with regard to its effect on virility.

---

[1] Op. cit., p. 240.
[2] Op. cit., p. 285.
[3] Beitrag zur Frage der Gesundheitsschädlichkeit der Salicylsäure. Med.-chir. Rundschau. Wien, 1887, Heft 14, p. 549.

Camphor, lupulin, antimony, niter, and the bromids are also said to have an unfavorable influence on sexual vigor. Krafft-Ebing[1] says in this respect, "Our nomenclature presents a large list of anaphrodisiacs, but when we practically try all these remedies we soon convince ourselves that they have no such virtue, or very little. This is true, for instance, of camphor, belladonna, and lupulin. Of somewhat more value are the bromids in large doses. No effect must be expected from doses of less than six grams.

"**Monobromated camphor** seems to be of quite exceptional value as an anaphrodisiac. Lupulin is not to be entirely disregarded; only it must be given in doses of over one gram each, if any effect is desired.

"Recently **antipyrin** has been employed as an anaphrodisiac, and in doses of two grams it is said to exert a sedative effect on the sexual apparatus. Hammond and a few others direct our attention to the anaphrodisiac effect of **sodium nitrate**. Quite recently I have treated painful sexual excitation with doses of three grams of sodium nitrate *pro die*, and have succeeded in reducing it to a minimum thereby."

Here I may be allowed to state that in the first German edition of this work, which appeared in 1889, I directed attention to antipyrin as a possible anaphrodisiac, saying, "Experiments made on animals having proved that antipyrin has an irritating influence on the inhibitory reflex centers, it would be very interesting to examine its influence on erectility."

Van den Corput has ascertained that, besides salicylic

---

[1] Die Therapie der Geisteskrankheiten. Wiener med. Presse, 1891, No. 22.

acid, quinin, menthol, phenol—indeed, almost all antiseptics—diminish sexual vigor in a marked degree. He considers that this fact may have its explanation in the inhibitory influence which these substances exert on the formed elements of the blood and on the spermatic cells in the same manner as on the lower organisms. The microscope shows, moreover, that these substances render the zoosperms perfectly motionless. According to Van den Corput, the diminished sexual vigor is to be ascribed to anesthesia and paralysis of the centers which govern the sexual function, as well as to the sterilizing and antivital influence which the antiseptics have on the spermatozoa.

INHERITED PREDISPOSITION TO IMPOTENCE.

There is an apparent inconsistency in speaking of inherited impotence, and yet the physician meets with many cases of sexual weakness and abnormal conditions of the sexual desire which after closer examination he cannot but trace to inheritance as the original cause. Hoffmann[1] says, in this respect, "It is a fact that there are men who from their birth either lack the incitomotor impulse which dominates over the sexual functions, especially erectility, or in whom it appears abnormally impaired."

It is quite conceivable that such a condition can occur in an otherwise normal state, this condition having been called by the ancient Canonists, who were very expert in such matters, "Natura frigida."

Those individuals are indeed very scarce who display complete inertness of their sexual life, although possess-

---

[1] Lehrbuch der gerichtlichen Medicin. Wien, 1881, p. 53.

ing sexual organs quite normal in development and function.

Krafft-Ebing[1] says that individuals possessed of weak sexual power, in whom the lack of sexual instinct can be traced to the cerebrum, are very rarely met with, and are probably without exception degenerate beings in whom may be found other disturbances in the function of the cerebrum, psychically degenerative conditions, and even signs of structural deterioration.

There are numerous families the male members of which are conspicuously weak in sexualibus. These are not always sickly people, but now and then are robustly built and of healthful appearance; most of them have very light complexions and high-pitched voices, and very often show no other weakness than that of the sexual organs and functions.

Sometimes we may see a man whose father was sexually very powerful, but who was begotten at a time when the father's virility was already on the decline,— *i.e.*, when he was partially impotent in consequence of unwise management of the sexual power, though it may have been originally great. Again, we see whole families in whom education has implanted principles that will ever be an obstacle to the proper development of the sexual instinct, and thus, indirectly, of sexual vigor.

Circumstances like the above will not surprise any one who is acquainted with the laws of transmission by inheritance, and who knows that, besides forms, qualities and habits also may be inherited. Even recollections are said to be thus transmitted, this assertion coming from a competent source, Exner. As fertility,

---

[1] Psychopathia sexualis. Stuttgart, 1886, p. 29.

for instance, and the early or late appearance of menstruation can be transmitted from mother to daughter, so also may a son receive from his father, by way of inheritance, sexual power or weakness.

Haeckel has attempted to formulate laws of inheritance, and, among other things, he says, "In all organisms with separated sex, the primary and secondary sex-characteristics are inherited one-sidedly,—*i.e.*, the male descendants resemble the father in the aggregate of the essential sexual characteristics, while the females resemble more the mother."

Just as there is a great difference observable in the sexual impulse in different nations, so there are very great differences to be noticed in the ardor with which the sexual instinct announces itself in different families. There are sundry links of connection whereby nature subdues the different degrees that might otherwise be too striking. These combinations of physiological and psychological phenomena are commonly called *the temperament*, and we all know that children have generally the temperament of one or the other or even of both the parents.

An innate sexual weakness shows itself frequently, yet not always, in the conformation of the genitals. Although the structure may be quite normal, they are nevertheless of an abnormal flabbiness and paleness. The erectile tissue is not very firm; the prepuce—if there is not a positive phimosis—is thin and moves with difficulty over the glans. Such individuals are generally unassuming, and make true and devoted husbands; nothing in them could incite them to act otherwise.

In most cases this weakness can be traced back to childhood, when **incontinentia urinæ** existed, and there is no denying the connection between sexual weakness and incontinence of urine. This connection Lallemand discovered with that perspicacity peculiar to him. I have always found that children suffering from incontinence of urine had unusually small genitals, and when I found an adult affected with the above disease, he invariably was either quite or almost impotent.

Men who suffer from congenital weakness of the sexual organs are not inclined to excesses in venery, as has already been stated; but with them the most severe consequences may be brought about by sexual indulgence to an extent not considered immoderate with others. In such persons insignificant excesses, or onanism practised for a short time, will result in frequent, and to a certain degree incurable, involuntary seminal losses together with a great enfeeblement of every sexual power.

The different grades of sexual weakness are innumerable, and whilst one man may scarcely show any signs at all, another exhibits from the very beginning the most decided sexual decrepitude. We shall therefore be right in asserting that the different degrees of sexual vigor, or of resistance against excess in venery, rest principally on hereditary differences; for sexual vigor or weakness is oftener inherited than is usually admitted.

The individual differences in sexual feelings and sensations are perfectly obvious. One individual may live for sexual enjoyment alone, all his actions keeping in view that one main object,—viz., sexual gratification, which he enjoys to a degree of ecstasy. Another may remain almost indifferent in regard to love and woman, may consider coitus a necessary evil, and while performing it may be thinking of some other affair.

You may meet with individuals who, with robust constitutions and well developed genitals, have from their youth shown comparatively little taste for sexual enjoyment,—individuals who are not easily tainted by the bad example of onanism, and who, later in life, exhibit a certain reserve in respect to the other sex. With some, and they are probably rare exceptions, this absence of sexual desire has even reached the point of disgust. Such men are shocked by a somewhat licentious expression; they are amazed at the excesses of others, and look upon love merely as the means of bringing forth children. This congenital disinclination for sexual pleasure is called **frigidity**, and may become an obstacle to virility, or it may render copulation possible only under specially favorable circumstances.

Krafft-Ebing[1] ranks this frigidity among the neuroses that have their seat in the brain, and calls it anæsthesia sexualis, absence of sexual instinct, which renders ineffectual every organic impulse starting from the organs of generation, as well as every fancy, every visual, auditory, and olfactory sensation that such individual may experience in this one direction.

Only congenital frigidity can prove a serious hindrance to the development of virility, whilst frigidity which is sometimes the result of a certain mode of education really yields to the first sexual desires that assert themselves positively; here principles, resolutions, and vows give way.

Again, there are individuals who, with vigorous constitutions, normal development of the genitals, and very energetic sexual desires, nevertheless become tempo-

---

[1] Op. cit., p. 25.

rarily impotent, where we can find no other cause than an inherited general or sexual **nervousness** which, at the given moment, either excites the inhibitory centers of erection to an abnormal activity or sets the nerve-centers of erection out of function. Persons of this category are mostly from families where cerebral and nervous diseases are hereditary; albeit Beard asserts that children of neurasthenic parents are generally unaffected in that direction. The children will probably remain healthy when parents suffer from acquired neurasthenia; but when this disease itself is congenital or has appeared in the place of some other hereditary disease of the brain or nerves, then there is no doubt that such conditions are transmitted from generation to generation. No one can assert that there are no neurasthenic children.

In general, the disposition to neuroses shows many varieties. There are persons who can make enormous exertions in mental, physical, and sexual matters without being affected by neuroses, whilst a high degree of neurasthenia will visit others after only a slight effort in these directions. Therefore the different injurious acts must be considered as bringing about the occasion for the disease, whilst the positive or effective and real cause must be looked for in the congenital predisposition to nervous diseases. Perfectly robust and vigorous persons may be affected by this predisposition in a very high degree. The special predisposition to neurasthenia may be so intense that moderate, nay, even infrequent, intercourse has an injurious effect.

A similar condition is observed in epileptics, though very seldom, it is true. I have known such an unfortunate one who had an attack after every ejaculation of sperm. Continued use of potassium bromid quelled the sexual activity and stopped the attacks; but whenever a pollution took place, it was followed by an epileptic attack, however large a dose of potassium bromid or sodium bromid had been taken. In such cases, which occur but rarely, castration may be suggested.

Of the limited number of forms of congenital impotence we have yet to mention one,—viz., **perverse sexual sensation.** This disease is generally congenital, the severe forms always; while the lighter forms may also be acquired. The subjects are generally persons affected by psychopathia, who can satisfy their sexual desire only in a peculiar manner. Such persons are not impotent in the true sense of the word, as erection is not lacking with them, but is often very vigorous, and yet they must be called impotent because they are not capable of performing coitus in the normal way.

Krafft-Ebing gives in his work, which we have already quoted, an exhaustive description of sexual psychopathy, and he classes the perverse sexual sensation (which he calls the paresthesia of sexual sensation) together with the sexual neuroses having their seat in the brain. Magnan[1] endeavored to classify the various forms of diseased sexual instinct, and, localizing them in the central nervous system, distinguishes four groups.

Keeping in view the object of this work, we shall be very brief in discussing this disease, which is both important and interesting to every physician. Numerous observations and investigations were required to lead to a knowledge of this form of disease; a study carried

---

[1] Des anomalies, des aberrations et des perversions sexuelles. Paris, 1885.

on during centuries was necessary to protect many an unfortunate being from punishment because of disease. Even in our day there is much to be learned before a correct opinion can be formed of many a case of this sort. On the other hand, we must guard against being misled by a false love for humanity. Society is in the right to protect itself against dangerous individuals, and it is justified in destroying persons who assuage their amorous longings in murder and other acts of cruelty.

Krafft-Ebing holds that, in paresthesia of sexual sensation, the spheres of sexual fancy are perversely accentuated by the association of feelings which otherwise would physiologico-psychologically awaken disgust, being accompanied by pleasurable sensations; this association may reach so high a degree as to become passion. The result will be perverted actions. This occurs the more readily when the pleasurable sensations, having reached the height of passion, inhibit or overpower adverse ideas with corresponding unpleasant sensations; also when these latter cannot be roused at all on account of lack or loss of the moral, esthetic, and righteous perceptions. I am of the opinion, however, that Krafft-Ebing goes too far when he says, farther on, "We must declare as perverse every manifestation of the sexual instinct which is not in accordance with the aims of nature,—*i.e.*, with propagation." At any rate, Krafft-Ebing does not mean to say that any and every copulation not undertaken for the purpose of propagation must be declared as a manifestation of a perverse sexual feeling; for then there would be few people in this world who were not, are not, or may not be affected by paresthesia of sexual sensation as thus defined.

If this perverse sexual sensation is congenital, it is in

most cases accompanied by particularly vehement or impetuous manifestations of the amorous desires. This abnormally increased sexual desire, which Krafft-Ebing terms hyperesthesis of the sexual sensation, is seldom a disease in the true sense of the word, and its explanation is, in my opinion, to be found rather in the circumstance or fact that abnormal sexual desires can be carried out only occasionally or with difficulty. Besides, the affected individual revolts, in proportion to his moral strength, against the satisfying of the ever-increasing lust,—say, for murder or other lusts incomprehensible to men whose volition or desires are in a normal condition. The individual is unable to control himself only when these perverse sexual sensations have reached the point when they must be called hyperesthetic. Then a crime is committed.

Persons thus affected procure satisfaction of their lusts by the most remarkable means. It is hardly possible to introduce order or a systematic classification into the sundry forms of this disease, because the longer we observe and investigate the more new forms present themselves. It might, however, be attempted to establish four groups of perverse sexual sensations, according as the satisfaction is looked for in perverse acts—

1. On persons of the other sex ;
2. On persons of one's own sex ;
3. On animals ;
4. On inanimate objects.

In the first group we should have to place first of all lust-murder and similar phenomena, as, for instance, different acts of cruelty to females. It is a fact long known that cruelty and voluptuousness are sometimes associates; a telling witness is the novel "Justine," by Marquis de

Sade. Here the monster pretends even to pose as a type with his perverse lusts. It seems to me extravagant, however, when Krafft-Ebing speaks of voluptuous kissing approaching biting in a chapter made up of lust-murder and allied phenomena. It certainly is only by individuals who are decidedly psychopathic that real cruelties are performed for satisfying sexual lusts, and the lust-murderers who use the knife and the dagger are all without exception suffering from mental aberration.

Several years ago I watched a case of this nature. The mother of a poor, fourteen-year-old boy, B., noticed that the body of her son was covered with black and blue spots, particularly the arms, buttocks, and thighs. After an examination, the boy confessed that his fifteen-year-old friend, P., son of an aristocratic and rich family, had induced him by gifts of money to allow himself to be pinched. When the little tormented fellow found the pains too great he began to cry and scream, whereupon his torturer commenced striking him with his right hand while he moved his left quickly to and fro in the left pocket of his trousers. When the cruel boy was afterward brought under my notice, I learned that he suffered from epileptic fits (epileptics are frequently subject to perverse sexual sensations). He was, on the whole, a well-behaved, peaceable, and talented child, but occasionally very disobedient, headstrong, and passionate. I convinced myself besides that he was an onanist. When alone with me he confessed that torturing his friend, whom he liked personally, afforded him a special delight, and that the ejaculation which he brought about at the same time was much more pleasurable than when caused by masturbation without his tormenting any one. The grandfather and an uncle on the mother's side died in an asylum; the mother was a sufferer from hysterical attacks. The father was known as a high liver; two brothers and sisters died at a tender age of what the mother called "Fraisen" (convulsions). I am of the opinion that the boy, whom I watched for some time, will meet with a sad fate in spite of all the care and treatment that he may be receiving.

The most common form of perverse sexual gratification is that which manifests itself in love for one's own sex. The men who love men and the women who love women, the "urnings" by birth, are probably but few, but more common are the pederasts and the priestesses of Lesbian love from necessity and lack of something better. The impossibility of satisfying sexual desire in the natural way in educational institutions, convents, prisons, on board ship, etc., leads many persons to such perverse acts. More frequent than is generally thought is pederasty between husband and wife, to prevent unwelcome progeny. Some persons practise it with prostitutes to avoid contagious diseases.

This is the kind of perverse sexual sensation which, more than others, can be and is acquired by people who are not predisposed by inheritance, but who, in consequence of blunted senses for natural charms, sink step by step, and finally find pleasure in loathsome and disgusting acts. Impotence often leads men to this vice, and hence it is generally old people who are addicted to it.

In the Orient pederasty is very common. Even Moses had to decree capital punishment for those guilty of this crime. Some authors think it is frequent there because of the fact that in Oriental women the genitals relax at an early age and become rather capacious.

If love for one's own sex has for its cause a congenital

perverse sexual sensation, the subject will usually show something peculiar in his character and appearance. His mode of thinking and of feeling is changed so as to correspond to that of the opposite sex. Most of these individuals whose perverted senses are due to inheritance betray their inclination by their conduct, often also by the garments they wear. The males have a liking for the occupations of females, and *vice versa*.

Some authors, as Gley and Magnan, endeavored to explain this phenomenon by assuming the presence of the sexual glands of a male associated with the brain of a female. Whatever the explanation may be, it is a proved fact that the perverse sexual sensation of an urning is absolutely independent of the will, and that many such persons have gone through the most dreadful struggles, but could not escape their fate.

Krafft-Ebing, basing his conclusions upon his experience, contends against the assertion of Tarnovsky, that a real urning—*i.e.*, one afflicted with congenital perversion of the sexual life—may, through education, be freed from his morbid sexual inclination and led to a normal sexual life. Krafft-Ebing admits that a good education will act here in the same manner as in a man with normal feeling but sensual, and will enable the individual to strive toward mastering the impulse, avoiding pederasty, and counterbalancing the desire, but only so long as this perverse desire does not assert itself with abnormal strength.

The third group of sufferers from perverse sexual sensation are the so-called Sodomites, people who gratify their lusts with animals. Sodomy was part of the religious cult of several ancient nations, including the Egyptians.

A man may by different causes be led to sodomy. Generally it is weak-minded persons, cretins, imbeciles, or idiotic people who, in their sexual excitement, have intercourse with animals. This excitement may occur periodically. In exceptional cases, also, persons apparently psychically sound may, for want of something better, abuse an animal if the occasion is offered and if they feel sexually excited. Very seldom is it moral degradation that induces a man or a woman to seek an animal for the gratification of amorous desires, but frequently elderly unmarried women use dogs for various unsavory purposes.

I had an opportunity to observe a case of sodomy. In a small provincial town a thirty-year-old man, an army officer, was caught in the act of gratifying his lust with a hen. One hen after another had perished in the house, and efforts were being made to discover the cause. When he was asked in court how it came that he had turned into a cock, the defendant suggested that the smallness of his genitals made intercourse with women impossible. An examination proved the assertion to be well grounded. The individual was psychically normal. Unfortunately, I neglected at the time to make a searching examination and to investigate the past history of the case, as the subject had not much attraction for me at that date (1877).

The fourth and last group of perverse sexual sensation consists of those who satisfy their lusts on inanimate objects, and, of course, does not include the different kinds of onanism with manipulations on lifeless things. The observations in this direction are not always quite reliable. It may be mentioned merely that in most cases articles of women's toilette, such as linen, nightcaps, shoes, etc., have occasionally the power to excite sexually and to satisfy individuals of perverse sexual feelings, who are, as a rule, by inheritance predisposed to mental diseases. Krafft-Ebing[1] says, "In other cases the sexual desire is roused by the sight of a woman's underwear, and is satisfied by their manipulation." If the center of ejaculation is in a state of irritable weakness, the mere putting on of such clothes suffices; otherwise masturbation must lend its help. Again, some individuals have to tear these articles to pieces in order to cause ejaculation.

Tarnovsky speaks of a psychopathic individual finding sexual gratification in the manipulation of peltry. Again, isolated cases have been observed where statues were appealed to for sexual gratification. Very instructive are a number of cases reported by Albert Moll.[2]

Ghastly and horrible is the defilement of corpses, of which even as ancient an author as Herodotus has spoken. One cannot easily imagine that any person in sound mind could be capable of such an act.

And now, at the conclusion of our discussion on the inherited forms of impotence, we have to mention some isolated causes of psychopathic conditions by which the sufferers may be rendered temporarily or permanently impotent, or at least sexually weak. In the first instance, idiots possess very feebly active sexual life, which is wanting entirely in idiocy of high grade. Sometimes the sexual instinct appears periodically; but then it is of a very violent character, and the idiot, resembling

---

[1] Op. cit., p. 48.
[2] Untersuchungen über die Libido sexualis. Berlin, 1897.

then a wild beast, seizes without warning the nearest female, even though she be his own mother.

Finally, for the sake of completeness, we should note that impotence and heightened sexual impulse accompany some mental diseases.

NEURASTHENIC IMPOTENCE.

Under this collective name we shall discuss all the forms of impotence dependent on a gradual degeneration of the sexual nerves and their centers. The present state of science does not disclose the character of this degeneration.

Before Beard every symptom of general neurasthenia used to be named and described as a distinct disease. In like manner the symptoms of neurasthenic impotence were described as special forms of disease and as special forms of impotence. People formerly spoke, and still do speak, of irritable weakness, psychical and relative impotence. Now, as the name "neurasthenia" has become so fashionable, and as the neuro-pathologists employ this collective name for forms of diseases that were hitherto designated by distinct names, we shall follow their example and describe under the name of neurasthenic impotence all forms of sexual weakness the origin of which we cannot trace to any structural change in the organs of erection and secretion, or which cannot be ascribed to any distinct appearance of disease in the body or in the so-called psyche.

Time will answer the question whether by this change we shall gain any benefit for the therapeutics of these diseases. At any rate, a clear distinction between those forms of impotence that arise merely from an affection of the nerves or nervous centers of the sexual apparatus and the other forms of impotence must be of great importance in the development of therapeutics.

Beard's fame has disturbed the peace of mind of our authors. Every day a new name is invented and dished up for the already perplexed practitioner. Every author has a new-fledged name for the old phenomena of disease, and our hope must be that these new names will not all live to an old age.

Impotence as a consequence of sexual neurasthenia either has its origin in a congenital predisposition or else is acquired. We have already spoken of the congenital forms, and shall now give our attention mainly to the acquired forms of sexual neurasthenia.

Neurasthenic impotence is less often caused by disease than by bad management of the natural sexual power,—*i.e.*, by **excess in venery,** either for the time being or habitually, by onanism, and occasionally by abstinence of longer or shorter duration.

Every man's virility has its limits, and these must be respected, for daily experience teaches us that impotence is most frequently caused by abuse. At the same time we must insist that the expression " excess in venery" has been misapplied without limit by most of our authors. To form a correct opinion in individual cases, and to distinguish between sexual excess and normal though frequent use, one must, first of all, possess personal experience in the matter, and must have observed a good number of *bon vivants*. Even the most extensive experience will not, in all cases, enable us to decide whether or not excess has taken place. Lallemand says,[1] "I call abuse every abnormal use of anything. Concerning

---

[1] Des pertes séminales, tome i. p. 315.

the generative organs, I understand abuse to be every irregular, premature, or other action which cannot result in the propagation of the race. There are no doubt many connecting links between these abuses and sexual excesses."

Lallemand, therefore, distinguishes between an abuse of the sexual power and excess in venery, and we shall retain this very sound distinction, as it will afford a special clearness to the subject under discussion. This is the more desirable because the two expressions thus separated to define very different things have usually been confounded and made the object of the most diverse and remarkable views.

Suppose an individual to have enjoyed coition without exceeding his natural power and without having had either the desire or the possibility for propagation, then, strictly speaking, he may be said to have committed an abuse of his sexual power, but certainly not an excess. I believe that a medical man need not concern himself about the abuse of virility in a natural manner. Let each be his own judge, and if any one requires a judge, let him address himself to his confessor.

If we choose to call an abuse of the sexual power every act of coitus that has not been undertaken with the possibility or even the intention of propagation, then abuse of virility will be carried on as long as there is a sensible and virile man in the world, and we need not feel alarmed about it. If coition, however, were to be accomplished only when a woman is to be impregnated, then most men would become impotent from continence, and a great many would become insane. My conviction is that the physician is concerned with excesses in venery only when they injure the body.

Lallemand[1] defines excess in venery in these few words: "L'usage poussé au-delà des besoins réels." This definition, however, admits of more than one explanation, since the meaning "real wants" is indefinite. Is a man actually to wait until the sexual instinct awakens without any action of his own? If so, many men engaged in serious occupation would never come to sexual gratification. The pleasures of love would then hardly be reserved for any one except the man of leisure and the idler.

I think a better definition would be this: Excess is coition for which an *effort* is required. Coitus easily performed and for which the individual does not require long preparation can never be called excess, even if no "real want" is to be satisfied. We eat and drink very often without feeling imperious hunger or thirst, and yet no one would pretend that excess has been committed. It is simply meal-time, and we do not feel the least dislike for the viands that have such an agreeable odor and a still more pleasing taste.

For the explanation of the specially injurious effects of sexual intercourse for which more or less effort is required, we must look to those cases where copulation or even excess is attempted or accomplished with uncongenial, nay, repulsive, mates. It is simply incomprehensible that Deslandes and Hunter could say that copulation with a woman, not rousing any special feeling in the man, is not so hurtful as when passionate love accompanies the act. Every day's experience convinces us of the contrary. Men who, for some reason or other, give too frequent proofs of very ardent love to

---

[1] Op. cit., p. 489.

a woman really disliked, suffer comparatively much more from it than those who give to a beloved or, at least, sympathetic being proofs of their love by still more frequent embraces.

It must also be observed here that those forms of impotence that arise from sexual excess, and particularly from too frequent coition, are mostly of a slighter and transient character, and comparatively easy to cure. Only after protracted abuse may those conditions occur that lead to real and sometimes incurable impotence, the so called paralytic impotence.

Man, as an animal gifted with reason, has reached special excellence in various spheres. In many a province of knowledge and art he has left marks of his inventive genius. In most things it took him thousands of years to pry into and lay open the secrets of nature. Some fields are still left fallow by his investigating power; he has never felt a desire to turn his attention toward them. He possesses quite *perfect knowledge* of the means *of abuse of his sexual power*, and he knows how to commit all kinds of sexual excesses. This knowledge he acquired at so early a date that even the very oldest of monuments of human culture tell of it as of something that had existed since the most remote antiquity. Modern culture, in spite of all the lamentations of tiresome, old, or hypocritical moralists, is not to be blamed for the almost universal abuse of the sexual power. This has been transmitted to us, like many other detrimental inheritances, by preceding ages, and our time suffers under it neither more nor less than it does from many other bad qualities, perverse notions, and unsuitable institutions handed down to us by our ancestors. Moreover, we find among uncultivated nations the most remarkable and cunningly invented sexual perversities. The charm-rings in use among some of our arrant rakes are mere toys in comparison with the socalled "ampallang"[1] of a few savage tribes. The only merit that can be claimed by our present time is that we have laid bare and shown in its true light the weakness of man in committing excesses in venery, and that we have exerted ourselves in discovering remedies for checking this devastating evil.

I am convinced that the united efforts of all the better elements of our society will succeed in preserving the nobler part of it from excesses in sexual gratification; and, though I do not expect to see an end of the excesses in venery, I hope to see onanism curtailed in the number of its victims, though it may be only in the better circles of society. This latter expression, "better circles," I wish, however, not to be understood in the sense commonly attached to it nowadays.

In general, and leaving out of question the perverse manifestations of amorous desire of which we have already spoken, it is found that excesses in sexual enjoyment may take place in two distinct ways. There is excess in copulation, and that of self-abuse or onanism.

It is hard to determine at which point of natural coition normal indulgence ends and abuse commences; and to fix on a certain frequency of sexual intercourse seems merely ridiculous. Regarding the frequency of copulation to be allowed we shall not easily agree; for, as there was a Martin Luther who allowed two conjugal acts per week, we have also an Acton who will allow one per week, and a queen of Aragon who demanded six per

---

[1] Mantegazza, Gli amori degli uomini. Milano, 1868, vol. i. p. 108.

day. The Talmud prescribes one act of coitus per day for a man in comfortable circumstances, who is strong and has no heavy work to perform. It allows two per week for a mechanic, and only one per week for scholars and laborers.

In the consideration of this question there must be kept prominently in view the great and various differences in individuals. Just as the digestive power differs in different men, and the act of thinking is not of the same rapidity in all, so also is the sexual capacity of very different grades. The appetite of one person may be perfectly satisfied by a small quantity of food, anything beyond that causing discomfort or nausea. Another person digests much larger quantities of food without inconvenience. A third may, after coitus, be disabled for a fortnight, whilst a fourth, after coition repeated several times at short intervals, can hardly await the next happy hour of love. I do not think I am in error in making the assertion that there is no excess as long as no unusual effort is required, no special means is made use of for rousing sexual passion, and no feeling of fatigue or faintness is experienced, regardless of the number of copulations, even if repeated at short intervals.

In the determination of sexual excess, a despicable, narrow-minded way of moralizing has come down from author to author, these gentlemen seeming to have forgotten that they are of the medical and not of the clerical staff. Such hypocritical sermonizing will certainly convert no one. As long as there are lovers there will be excesses in the estimation of the authors, but not resented by nature. Nature punishes only such as act contrary to her laws by employing various means to rouse desires and to irritate the strained nerves to immoderate activity.

Since virility is not of the same degree in all individuals, the *limit between normal use and excess* in them must also be different. Without regard to age or the state of health, the disposition at the time being is of extraordinary influence. That which was moderate indulgence ten years ago may be excessive to-day; and even what might have been accomplished easily and without fatigue a few days or a few weeks ago may be hurtful now. So also what may have seemed quite normal with one woman may have to be considered as a decided excess if committed with another woman.

Inquiring into the *reasons* that induce a man *to commit sexual excesses*, we cannot, after what has been said above, accuse age, amativeness, temperament, etc., as Lallemand does; for such manifestations of the sexual instinct as are produced by youth, temperament, or sensuousness cannot be called excesses. The inducements to excess in venery are only few; in the order of their frequency they are: masculine vanity, love, sensual women, different conditions of irritability of the sexual organs, and hyperesthesia of the sexual centra, which may pass into maniacal conditions and become a veritable satyriasis with priapismus.

Masculine vanity is the commonest cause of sexual excesses, as Lallemand[1] has already observed. In wedlock and out of wedlock the man feels the desire to impress the woman with his power. He starts from a principle that is correct in itself, but he very often carries it out in a faulty manner. Over-exertion in sexual mat-

---

[1] Op. cit., p. 614.

ters seldom impresses the woman favorably, and inevitably leads to humiliation; and besides, the descent from the pedestal of a hero often leads to dissatisfaction in marriage, and may induce the wife to seek elsewhere that which she has learned to like and which her husband can afford her no longer.

Love may sometimes give occasion to excess. The desire to unite with the object of love as often as possible may overtax the sexual organs. Love is decidedly a powerful stimulus, but, like all stimuli, it loses its power in time. Love may have lost its stimulus, but may still live if the cause lies only in the giving way of virility. When love, however, has lost its stimulating power, although the man has retained his vigor, then love has simply ceased to be love.

Voluptuous women may now and then be the cause of sexual excess, but it must be said that in general oversensual women are the exception, and novels and anecdotes know more about them than does actual life. Almost without exception, woman grows unduly sensual only in commerce with sensual men; but if once she is so, then woe to the man whom she holds under her sway, if he has not the courage to wish her good-by in time.

Occasionally skin-diseases about the genitals cause too frequent erections by external irritation. If we except gonorrhea and its sequelæ, quite as unfrequently there may occur irritation of the verumontanum or in its vicinity, causing untimely erections and thus provoking excesses.

Some persons have naturally the disposition to love always and much; they do not feel happy unless they are in love, and when they are they commit so-called excesses, which in reality are not to be so regarded, but which should rather be considered as manifestations of their sexual power. Indeed, we see that such fortunately disposed natures receive no harm whatever from these apparent love-extravagances, as they endure mental and physical exertions exceedingly well. Such natures are oftener met with in southern countries than in northern, oftener belonging to the better educated classes than to the uncultivated. They are mostly of a temperament easily roused, generally superficial, and in common life are called light-minded people, but sometimes deserve a better name, since they accept life as it is. As a rule, puberty shows itself in such persons very early, long before it appears in other individuals of the same nationality. Mantegazza[1] says, "To precocious puberty correspond luxury, polygamy, and libertinage." The coincidence of early puberty and excesses in southern nations, may find its explanation in the fact that persons entering upon the possession of their sexual power before they possess full responsibility are more apt to commit sexual extravagances.

Again, *abundance of power* will be sure to *lead to excesses*, as authors understand them; but these, in my opinion, are not excesses so long as the right proportion with the force is maintained. Decidedly enviable are those so fortunately gifted by nature, whom we might call with Mantegazza[2] "grandi amatori" (great lovers), and of whom he says, "The great lovers are frequently weary, but in their weariness there is not a shadow of ennui."

---

[1] Gli amori degli uomini, vol. ii. p. 233.
[2] Fisiologia dell'amore. Milano, 1882, p. 383.

Excesses in venery are generally committed by men of a noble character, who subordinate hygienic considerations to the pleasure of sharing the highest delight with a beloved creature. Hence it is generally genial natures endowed with artistic talents that worship sensual pleasures and often fall a victim to them; whilst egotistical and mean natures calculate in numbers and are stopped in love's intoxication by the thought, "That might hurt me." It is, however, very strange that those individuals that take so little care for the preservation of their virility are often the very ones who keep it until an advanced age, whilst those who have always husbanded their power so economically often lose it prematurely. The force of habit plays here an important rôle. Some descend from much to little, and others from little to naught.

I beg leave here to advance an opinion which is in opposition to the views of most authors,—viz., a strong and healthy man may, by means of reasonably active gymnastics of his sexual power, increase it considerably without damaging his health, since a vigorous metabolism is capable of rapidly replacing all such losses. I should hesitate to express this opinion in a book subject to the chance of being read and misunderstood by laymen; but, in a work written for medical men exclusively, I feel it my duty to state the bare facts unadorned as experience has shown them to me, and thereby obtain the result, perhaps, that some doubt may introduce itself into the realm of common petty ideas of narrow minds.

A gland that has seldom been excited to ordinary, and never to energetic activity, will never yield the same quantity of secretion as one that is never overtaxed, but is roused frequently and vigorously. Exactly the same may be said of the muscles of the apparatus in question, which, by reasonable use, may be strengthened and rendered capable of performing their functions. The glands and muscles active in sexual intercourse, all of which are secreting or at work through the influence of the nerves in action, are subject to the same laws as the nerves themselves, and will consequently be assisted in their secretion and strengthened in their activity by moderate and even stronger irritation, though they would be weakened by too strong an irritation.

It is certain that the nervous system is strengthened by every inducement to action as long as the action required is within the limits of performance without effort. Too much, however, is here more damaging than too little, and the happy mean is to be followed.

Too little irritation is apt to render the nerves and their centers inert in reaction and weak in functional power; thus, the genital glands which they innervate will be made indolent in secretion and the muscles of erection weakened and stunted. Again, too much irritation affects badly the co-operating nerve-apparatus, and causes first a state of excitation and then of relaxation. This is the reason why men who are habitually excessive in venery are for a time specially powerful, for the sexual nerves are in a state of excitement; but this condition gradually changes into a state of weakness, which, in its turn, passes over into a state of paralysis if rest does not intervene in time. "Lasting immoderate irritation of the nerve, without a time of rest sufficient for recuperation, gives rise to lassitude at first, and then leads to diminution of the excitability by exhaustion of the nerve; and yet the nerves are possessed of extraordinary endurance with respect to the most various irritations."[1]

Decrease of the sexual power is for one who commits excesses in venery a sign, given by a sometimes kind nature, that it is time to retreat. Unfortunately, this sign, though noticed by most offenders, it is true, is wrongly interpreted, and may even cause impatience or anger. A struggle against nature begins; moral, physical, and even medicamental excitants are brought into service, but even these subsidiary forces leave the individual in the lurch, as the combat is an unequal one, and impotence is the result.

Now, instead of looking for help, the impotent man gives way to the general belief that there is no herb growing for impotence. The physician often supports this belief by sending the unfortunate patient away with a few well-meant phrases of solace, instead of proceeding first to a correct diagnosis and then to the determination of the appropriate treatment. Thus, then, the patient is generally left to himself; recourse is had to new and stronger excitants; he torments himself to bring about coition; these fruitless efforts aggravate the evil; they dull the nerves and their centers more and more, until, finally, they exhaust them, so that paralytic impotence results.

In order to comprehend the pernicious effect of habitual excesses in venery, we must first of all consider more closely the **after-effect of the sexual act**. According to most authors, man, after copulation, is in a state of exhaustion, the duration of which varies in the

---

[1] Landois, Lehrbuch der Physiologie. Wien und Leipzig, 1893, p. 673.

opinion of these authors. Now, generally speaking, this is not so. A robust man who is in possession of his full sexual power is, after coition, not exhausted or depressed at all. He keeps generally perfectly still, because the action of the heart and of the lungs has been accelerated and must now be moderated, and because he still holds the object of his love in his arms and ruminates upon the past happiness. Those individuals who feel exhausted after copulation had better not commence at all, and the authors who uphold the idea may possibly give expression to their own experience. With a man perfectly vigorous sexually the Latin phrase, "Lætc venire Venus, tristis abire solet," applies only to the same extent as in the case of a gourmand who enjoys a good appetite, and thus regrets, after a meal, that it is not to commence. The discovery made by Beard, that the pernicious consequences of coition often do not show themselves until the third day, must be characterized as at least peculiar. This discovery our American neuropathologist claims to have made on a patient who was unquestionably neuropathic, and, moreover, suffering from spermatorrhea. A man who feels the effects of copulation three days afterward ought to be forbidden sexual intercourse.

Now, although I deny that coition must of necessity be followed by a state of exhaustion, yet I must grant that coition implies or necessitates the *spending of a certain amount of force*. In the first instance it means a bodily exertion; secondly, the work performed by the nerves and their centers is considerable; and, thirdly, on the part of the male the loss of substance must be taken into account. It is certainly not the *bodily exertion* that causes the pernicious consequences of immoderate copulation, for we know that all bodily exertion is rather beneficial to the health of man. Is it the undue effort of the nervous system, or is it the excessively great loss of substance? Certainly neither the one nor the other alone, but both combined, with the co-operation of other influences about which we cannot yet say much.

To prove that the excessive loss of sperm causes the bad effects of sexual excesses, a *comparison of these effects on the two sexes* has often been made. It has been asserted that women suffer more rarely than men from the evil consequences of sexual excess. This assertion is not in accordance with daily experience, which teaches us that sexual excesses have on women a more serious and more lasting effect than on men. We need only to observe very young married couples. The recollection of every physician will show him the pale and fatigued face of the wife beside the comparatively fresh, though somewhat emaciated, face of her husband.

Attention must, however, be called to the fact that women very soon accustom themselves to sexual exertion, and then even excesses are more easily endured by them. Besides, if the woman feels a disinclination for a man, she can, without participating in the act, simply endure it. If disinclination on the part of the man can cause in him relative impotence, disinclination experienced by the woman may produce a kind of frigidity, so that while giving herself up to the act she can remain perfectly indifferent. Here we have, then, the key to the assertion of some men who declare women in general to be sexually greedy; whilst others—especially scholars, for whom the ladies have often no liking—declare them cold and insensible.

The *loss of substance* certainly plays a rôle of some

importance, and it is not only the circumstances preceding and accompanying the ejaculation, and certain processes going on in the nervous system [1] which cause the evil effects of sexual excesses. This may be understood when we consider that evil consequences of a greater magnitude follow the loss of sperm from spermatorrhea than from sexual excesses. In some cases of spermatorrhea of a high degree the loss of semen is unattended with consciousness, and we can hardly imagine any influence that could primarily affect the nervous system.

Paget[2] thinks that the morbid condition in the nervous system, and more especially in the spinal cord, caused by excessive coition is analogous to the condition which is observed in muscles after excessive efforts. The comparison between muscle-atrophy and paraplegia might be very instructive if sexual excesses ever caused paraplegia.

In this comparison by Paget of muscle-action and nerve-action I consider only the following observation correct, in which he says, "I cannot explain to myself why excessive coition infallibly causes loss of sexual vigor with certain persons, while the same excess causes in other persons no manner of disturbance. But the same different effects are observed in respect to muscle-effort, and remain also without explanation. A given amount of muscle-exertion that exhausts one individual and leads to muscle-atrophy develops the power in

---

[1] Curschmann, Die functionellen Störungen der männlichen Genitalien. Handbuch der speciellen Pathologie und Therapie von Ziemssen, Band ix., Hälfte 2, p. 476.

[2] Acton, Fonctions et désordres des organes de la génération, p. 237.

another, and increases the capacity and vital energy of the muscles at work."

What are the real **consequences** of immoderate coition? As far as my knowledge reaches, there is only one really serious effect, but that brings in its train a series of other evils: it is sexual impotence.

No one doubts that excesses in venery cause impotence. Those who have not denied themselves the enjoyments of life know that after such excesses there occurs a time during which desires lie dormant. With weaker natures such a time follows sometimes after very few excesses; even after a single night of revelling, there ensues a longer or shorter period when coition would be an impossibility. This condition cannot be called impotence, for it is *physiological*, similar to the state of fatigue, and is caused by the spending of the provision of sperm and the exhaustion of the sexual nerves and their centers. With sexually weak men this state lasts longer than with stronger ones. It will last longer after repeated excesses, and these pauses growing ever longer pass sooner or later into a permanent inability, which we justly call impotence. These pauses are certainly a wise provision of nature; they are, in a measure, bars placed in front of the impetuosity of youth. Without these bars there would be many more impotent persons than there are.

After sexual excesses the spinal cord is no doubt in a state of hyperemia. Hence the state of irritation in the centers of erection that continues after the excess, and the fatigue of the overtaxed nerve-substance which is plainly seen after the cessation of the hyperemia. After many such states of lassitude, which may pass away more or less rapidly, there ensues a state of exhaustion which may be permanent. This constitutes impotence. Impotence following sexual excesses is in the great majority of cases a state of exhaustion of the nerves and nerve-centers concerned,—a kind of sexual neurasthenia of different grades.

In the main, excesses in venery cause only sexual neurasthenia; whilst onanism causes, besides sexual neurasthenia, derangement of the organs of secretion, and especially of those of ejaculation. From this arise primary morbid pollutions, which finally produce impotence. The same is true of continence, but in rare cases only.

Neurasthenia sexualis arising from sexual excesses shows, like every other form of neurasthenia, quite peculiar phenomena, and its symptomatology resembles that of general neurasthenia in many points. As already stated, there are to be seen signs of exhaustion of different grades. Sometimes the desires of the patient are in marked contrast with his force. At times positive satyriasis may be observed in completely impotent men. Such individuals practise mental and actual onanism, as they are no longer capable of performing the sexual acts so ardently wished for.

Again, the sexual apparatus may still be comparatively apt in function, but the subject is nevertheless impotent because of a sexual **frigidity** which is not an unusual consequence of sexual excesses. The feeling of satiety which is observable after nearly every sexual excess, but which generally disappears rapidly, becomes permanent in some cases, and the individual, who still has erections now and then, is nevertheless impotent, as he feels an apathy toward the female sex. This state, it is true, is commonly observed only in persons who have never had a well-developed sexual instinct, and who, in spite of this, allowed themselves to be incited to excesses. The same state may, however, appear in persons of a strongly developed sexual impulse, who have committed repeated excesses with persons whose physical qualities were not worth such an effort. Such frigidity is particularly noticed after manifestations of power performed with a view to pecuniary gain, and in husbands who sometimes perform their conjugal duties for years, and conscientiously, but with reluctance.

Another form of impotence is noticed in the track of excesses in venery continued for years, in which the sexual organs may continue to be capable of function. After years of excesses there arises a state of **satiety for ordinary sexual pleasures,** a torpor of the sexual emotions for normal and natural satisfaction. A perverse sexual sensation develops itself quite gradually. The patient is incapable of performing the sexual act in a natural manner, but can still indulge in pederasty or other wayward acts to satisfy his perverse lusts. This form of impotence in consequence of sexual excesses is oftenest observed in old men, the majority of whom have some psychical affection. All the perverse actions of sick and also of insane persons are merely an extraordinary augmentation of normal phenomena, emotions, and acts arising from psychological and physiological origin; and similarly we can trace, in the forms of perverse feeling, a highly abnormal exaggeration of phenomena observed in healthy persons.

It is generally known, though not so generally admitted, that the original taste in sexual matters undergoes manifold changes with increasing age and after numerous sexual enjoyments. As there is a great difference in the individuals, it is impossible to establish a general rule or law, and I shall therefore attempt only to represent the process in the manner in which it generally presents itself for observation.

In the first place, we must entirely exclude the early years of ardent desires that come before completely developed sexual maturity. These are the years of sexual blockheadedness and awkwardness; the individual is without taste, and not at all particular in the choice of altars on which to make his sacrifices to Venus, although he dreams of ideals. When, however, that sexual bulimia, as we may call it, is over, taste appears, and the individual grows particular within the limits of his taste. This taste is good or bad in the opinion of his fellow-men, but, judged from his own stand-point, it is always good. The taste remains then generally settled during the years of the greatest virile power; but when riper manhood approaches it begins to deteriorate. The individual grows less and less particular, and will carry on commerce with persons he would formerly have rejected with contempt. This corruption of taste is the reason why elderly gentlemen often associate, for the time being, with slovenly and unattractive servant-girls. Closer examination will prove the error of the generally accepted opinion that elderly gentlemen are dainty in their choice, and that only the want of a better chance can make them less particular. Of course, due consideration is to be paid to the rôle played in this respect by education, the different conditions of wealth, rank, and other circumstances.

The abnormal exaggeration of this physiological degradation, as I would call it, in the sexual taste, is often the origin of perverse sexual sensation. The class of individuals who have thus become perverse comprises all those who, after a long, honest, and exemplary existence, come, in their old age, into conflict with the law, as pederasts, exhibitionists, or ravishers.

In a few isolated cases, and especially with sexually weak individuals, continued excesses lead to **paralytic impotence,** as already stated. This is equivalent to complete paralysis of the sexual nerves and their centers,—a condition which certainly occurs very seldom in young persons.

Besides the states already described, excessive venery causes transient feebleness in the entire body and disturbances in the functions of single organs. After a real excess in venery the individual feels weary and exhausted, may be nervously agitated for a time, and may be visited by vertigo and fainting fits. However, all these symptoms generally vanish after a few hours' undisturbed sleep following a strengthening meal. The individual will experience no further consequences.

I never saw or heard of ejaculations mixed with blood; although Lausac[1] has recently asserted that the causes of it may be urethritis, prostatitis, epididymitis, also sexual excesses, or long continence. I believe, however, that this is probably of exceedingly rare occurrence, and will hardly appear in normal individuals. I do not think that gonorrhea can arise from protracted coitus or sexual excesses, as Lallemand seems to believe,[2] if the subject is in perfect health. Nor can I believe that sexual ex-

---

[1] Untersuchungen über die Hämatospermie. Med.-chir. Rundschau. Wien, 1888, Heft 3, p. 95.

[2] Des pertes séminales. Paris, 1836, tome i. p. 586.

cesses can cause degeneration, suppuration, or induration of the cerebellum.[1]

Some individuals committing frequent excesses in venery, if they are strong, accustom their nature to these excesses after a time. Such wanton persons may commit great excesses almost with impunity for a long time. They consider themselves heroes in sexual matters; but, nevertheless, they steadily lose flesh in spite of an appetite that may be quite excellent. Exceptionally it is noticed that sexual excesses seem to favor the development of obesity. I have known a man, thirty-two years old, who was continually losing weight during continence and a state of good health, whilst he gained in weight when he was doing his utmost sexually.

Sooner or later the sexual power is sure to diminish noticeably after sexual excesses, even with individuals of great resisting power, and temporary impotence will then for a time prevent further excesses. This dallying with one's sexual power I have observed frequently, and have received many reports of the same kind from my patients.

I never saw the fearful consequences of excesses as they are described in books. I never saw one case of frequent pollutions or spermatorrhea caused by excess in venery. Considering that I have observed a great number of such wanton persons, and a still greater number of patients suffering from spermatorrhea, I may be justified in the assertion that immoderate coition without onanism cannot cause spermatorrhea. Among Oriental nations, where polygamy is the rule, and occasions for excesses are constantly presented, premature

---

[1] Black, On the Functional Diseases of the Urinary and Reproductive Organs. London, 1875, p. 112.

impotence is often noticed, but spermatorrhœa very seldom. Abnormal pollutions and spermatorrhœa arise now and again in the train of impotence, but never directly after excesses in venery, however long these may have been practised. This is the chief difference between the effects of onanism and those of excessive coition.

The consequence of excess in venery is always and without exception impotence, this impotence putting a stop to further excess.

It is difficult to obtain a clear idea of the manner of action of the pernicious effects of continued sexual excesses, because we are denied the possibility of watching the pathological changes caused by these excesses. Virility enfeebled by habitual excess does not make itself felt by any kind of trouble in the sexual organs, and sometimes, though no change in them is visible, the individual is impotent.

Although the rest of the body may be quite normal in its functions, we see that some, though not all, persons suffering from sexual debility lose weight. This degeneration of the forces, this general sinking, can be explained only by the injurious effect on the mind produced by consciousness of impotence; whilst the paralysis of the sexual nerves can be attributed only to the consumption of the nerve-power. However, we cannot say what this consumption really means. Whilst the question has scarcely been studied at all, and lies in great darkness before us, we can only say, *Continued excesses in venery lead sooner or later to impotence, with a rapidity varying in proportion with individual differences; and impotence brings very often in its train other unpleasant consequences.* If some authors state that there are men who commit excesses in venery with impunity, we must remember that what these men have committed were not excesses in the true sense of the word.

It is amusing to read the statement that some cases have been noticed where single excesses in venery have caused death. No doubt long-continued sexual excesses may cause temporary disturbances in the functions of other bodily organs, but permanent derangement certainly can occur only in weak and sickly individuals, since the sexual organs refuse obedience in proper time.

The assumption that sexual excesses may directly cause atrophy of the testicles, for example, is certainly erroneous, since impotence is always primary; while atrophy of the testicles is only secondary in consequence of impotence.

It used to be the custom to impute everything to sexual excess: a death whose cause was not clearly discernible was attributed to sexual excesses; if some one became insane, sexual excesses were blamed for it. Even in case the father and mother had been insane, inheritance had nothing to do with it. If a man died suddenly in an ill-famed house, sexual excesses were accused, and one forgot that the man had quite as good a chance of dying suddenly in his conjugal bed, though he rarely may have made the proper use of his conjugal rights. If any one should wish to convince himself of the exaggeration that is going on in these matters even nowadays, he has only to take up a modern book, such as Bourgois,[1] for instance, where he can read literally, "Dissolute persons are so frequently subject to spinal diseases that the special name of dorsal consumption or

---

[1] Les passions dans leurs rapports avec la santé et les maladies. Paris, 1877, p. 144.

phthisis, tabes dorsalis, has been given to them when they originate from sexual excesses." Dr. J. A. Spalding, of Portland, Maine, goes even further, and relates the history of a group of four cases of optic atrophy following sexual excess. All four subjects gradually lost useful vision in spite of treatment, though they did not become totally blind. As no other mortal ever had the chance to observe such consequences of sexual excesses, and Dr. Spalding, on the contrary, found four such phenomenal cases, we have to look toward the Portland climate or imagination for an explanation.

In no other disease has cause so frequently been mistaken for effect, and *vice versa*. Authors too often ascribe many pathological states to sexual excess, without noticing that these morbid states, together with their supposed cause, have a common and deeper origin in an inherited pathological predisposition. Thus, then, the effect was mistaken for the cause.

Generally speaking, real excess in venery seldom occurs, and is not often the cause of impotence or of a general physical degeneration. In case extraneous circumstances do not prevent an individual from accomplishing his own ruin by sexual indulgence, nature, as a rule, will refuse him the means for continuing his excesses. Not all women will consent to excesses; again, the individual, if somewhat energetic, may restrain himself from further excesses.

We shall meet with altogether different conditions when we make **excessive onanism** the subject of our discussion. If onanism causes degradation of the physical force and impotence much more frequently than excesses in venery, we find the explanation in this difference of all the conditions.

It is only an exception when excesses in venery can be practised before the individual has become pubescent, for he generally lacks the means and occasion therefor; onanism, on the contrary, may be practised by an undeveloped child, as, unfortunately, is frequently the case. Of course, every sexual excess committed by an individual before puberty is ever so much more far-reaching in its consequences.

Nature is seldom able to hinder the onanist in his destructive work until it is too late. The onanist wants no erection: he can without it bring about orgasm and ejaculation. The onanist need not wait for time of leisure or for a special place or occasion. He can satisfy his desire in bed under the warm covers, almost under the eyes of his parents and teachers, in the closet, in any dark corner. Some have acquired such dexterity that they can ever add new injury to the weakened sexual organs, which are almost continuously in a state of irritation, by clever manipulations of the penis by the hand introduced into the pocket of the trousers. This can be done at school, at church, at the theatre, at balls and other entertainments, or during a walk in the street or a ride in a carriage.

The recovery from this condition is more difficult to accomplish, and the ability to desist from further injuring the sexual power is more difficult to acquire, in this case than in the case of excessive venery. The individual is not able to avoid seizing the opportunity to practise his vice. It often takes a long time for the onanist to get a clear understanding of the evil consequences of his actions; but when he does discover his error, a fearful struggle arises between the pernicious, almost overpowering habit and the enfeebled juvenile, who may be a mere child. This struggle is so hard and dreadful that even a vigorous and energetic man might succumb.

There is yet a weighty circumstance that renders excesses in onanism far more fatal than excesses in venery. The one who commits excesses in venery is ever enamored, feels seldom any remorse about the excess he has committed, but is rather inclined to look upon it as a triumph of his irresistibleness and power. He is satisfied with himself, and draws immeasurable joy from the inexhaustible treasury of love. He is most of the time happy, and his joyous mood contributes largely toward the preservation of a healthy condition of his body. The reverse of this is the case with the onanist: he is always in conflict and exceedingly dissatisfied with himself; he is ashamed of his doings, and regrets to commit what is commonly called a " vice." Hence the onanist is ill-humored and melancholic, this state of the mind having an influence which almost without exception produces an injurious effect on the functions of nearly all the bodily organs.

Finally, it must be stated that it is not at all settled whether coition and masturbation are equivalent acts. I doubt it very much, because I know that with individuals accustomed to excess in venery the commission of a single act of onanism leaves them in a weaker and more dejected state than a great excess in copulation. I think that, after all, the circumstances preceding and accompanying ejaculation, and, most of all, certain processes in the nervous system, must be of more consequence in onanism than in copulation, or, that these " circumstances and processes" require a greater effort to be brought about by onanism than by coition, and consequently produce greater fatigue.

Every medical man who thinks about this matter must ask himself whether those authors are right who call onanism a vice. I do not think they are; for if we take into consideration the circumstances under which one falls a prey to onanism, and, moreover, the almost insurmountable difficulties encountered in resisting or overcoming the bad habit, we unquestionably must come to the conclusion that onanism should with more justice be called *a disease*. In the same manner in which one may fall sick without any fault of his, he may fall into the grasp of onanism and be unable to tear himself away even with the greatest effort.

The most common form of onanism consists in moving with *manu propria* the prepuce backward and forward over the glans until ejaculation is obtained. Sometimes the onanist produces friction of the prepuce against another object, as, for instance, by lying on his stomach and rubbing against that upon which he rests, thus causing the prepuce to move over the glans; less frequently there is mutual onanism practised between two men. This kind of onanism occurs most frequently among persons affected with perverse sexual feelings, but who have not yet sunk to pederasty. Still less frequently there is mutual onanism between man and woman; yet it occurs often enough, and seems to become a regular expedient in matrimony in some countries, whereby the increase of the family is avoided. Lingua et labia are then often used in the place of the hands.

Onanism as a manner of satisfying the sexual instinct is very widely spread. It is practised by almost every young man when he becomes pubescent, so that one might be tempted to look upon onanism as a physiological act; and this so much the more when we realize how eagerly monkeys and other animals practise it.

Onanism is not without its vindicators; only they have not the courage to speak aloud and publicly in behalf of their belief. In a small circle of pupils I once heard a very popular professor say that onanism moderately practised has its advantages, particularly for students, as money and valuable time are saved, all unpleasant connections and obligations are avoided, no one is made unhappy, and there is no danger of contagious diseases. If considered from an egotistic stand-point, these reasons might be correct, if only the onanist were capable of keeping within moderate limits; but it is this excessive onanism that renders it so particularly hurtful. The onanist's resolutions of restraint may be compared to the " serments d'ivrogne." Just as the drunkard, sitting with the last of the number of glasses that he vowed should not be exceeded, continues to allow himself just one more, "always the last," so the onanist bargains with himself, and his "last time" becomes centuple, in spite of his vows, oaths, and promises.

Although we have really to deal with the bare facts only, we feel interested in the **causes** of this universal practice of onanism, because, while obtaining a knowledge of the cause, we may get a knowledge of the remedy.

There is no doubt that onanism is of very old standing, perhaps as old as the human race. Over the whole extent of human history there is no lack of statements, or at least hints, about onanism. Hebrews, Greeks, Romans, all were acquainted with it. It may be quite true that in olden times onanism was not so common, but civilization cannot be accountable for the difference; for in those times the young people did not lead sedentary lives; it was easier for them to procure natural sexual gratification, if they were not already pederasts. Pederasty, however, always goes side by side with onanism. Again, the intercourse with venal women was not dangerous, because there was no syphilis in those times.

If onanism has not already become a habit, it is out of the question as long as the sexual instinct can be satisfied in the only natural way, in the arms of a responsive woman. This is, besides, the manner more in accordance with feelings of love; it is the more pleasant, the nobler act. When, however, this mode is denied, then with virile men erections occur, and from the temptation of feeling and fumbling about the genitals there is no great distance to masturbation. The present social regulations cause ever increasing difficulties in the obtaining of a wife, and we should not be surprised to learn that onanism is really much more common at the present time than during any part of antiquity. Onanism is not practised by the monkey when living in liberty where he can go after his mate; nor by the bull nor the stallion that has enough of cows or mares brought to him; nor with the dog not led with a line by his master. Apes and other animals fall into onanism only when proper mates are not obtainable.

There are many writers who assert that the cause of the wider and wider spreading of onanism is to be found in the ever-increasing corruption of morals, or, as some like to consider, identical with enlightenment. General attention has been directed to onanism but recently. We will grant that onanism has been "ever spreading," though there is no proof of this, but we must repel the reproach laid at the door of modern culture or civilization. Modern culture has nothing to do with our bad social conditions and crazy views derived from medieval times which are the chief causes of onanism. These conditions and views came down to us from the time of the greatest religious intensity; and they remain in force, not because of our modern culture, but in spite of it. We can only hope that when this culture becomes common property it will free the world from these fetters that hinder free action, and then onanism will be shown its proper place. Certainly it cannot be denied that the present mode of education of our children offers many moments favorable to onanism; but this very mode, too, is an inheritance from the dark ages.

Children learn onanism mostly through *seduction*, and, as a rule, one child learns it from another. A single so-called black sheep often suffices to corrupt all the others in a family, institute, school, etc., whereby the frequency of onanism in boys' and girls' institutes finds an easy explanation. The younger a child is, the easier it is to allure it, even those who at first resist seduction finding later on a taste for it. It occurs, though seldom, that coarse, uneducated bad adults, even teachers, find a pleasure in misleading immature children to onanism. In the literature of the subject we read of cases quite incredible. Thus, there are nurses and nursery-maids who understand how to quiet screaming children by playing with and sucking the child's genitals. It is to be remarked that the little screamer is easily and quickly silenced, but thereby a state of irritation in the sexual organs is caused, and the child induced to pull and play with them, and finally to practise real onanism. This is the more injurious the younger and weaker the child is.

Rarely, *unwise parents*, in their anxiety for the child,

clothe their warnings in such awkward language that it draws the child's attention to sexual matters.

I regret to be obliged here to add that occasionally *the study of certain subjects* may lead to onanism, and, unfortunately, it is, strange to say, principally the instruction in a subject that is generally thought to further morality. In the sentences and passages to be memorized words are used which quite evidently have reference to sexual matters. Immature children are served with little stories of the most piquant contents, which are for them the first hints to sexual things. Dull children think no more about them, but sprightly and intellectual children will ponder over these legends. Dried-up old pedagogues are not aware of the kind of thoughts that present themselves to the child when meditating on certain stories. Then, again, we have in our high-schools the Latin and Greek classics with their often vulgar and blunt language. It is well known how eagerly and assiduously the passages of an erotic tendency are read and the greatest linguistic difficulties overcome, even if such passages are skipped by the professors.

Withal, however, I do not mean to say that a youth should be brought up in ignorance of all sexual things. I do think that, by all means, a child has the right to be exempted from obscene ideas as long as he is sexually unripe and his nature does not assert itself. However, when puberty has arrived, and has announced its presence by unmistakable signs, then I think the youth should be told all the truth, without allowing him to be excited by piquant reading. Youth may find that piquant which leaves a man perfectly indifferent.

What are we to say or to think of a medical man advising mothers to endeavor to accustom the prepuce of their boys to remain behind the corona glandis?[1] As this can be accomplished only by moving the prepuce repeatedly back over the glans, such mothers would, in plain words, masturbate their sons.

Bad example is the most frequent cause of the spreading of onanism, and causes the greatest injury, because it brings under the sway of onanism quite immature children who up to that time did not feel the least movement of sexual desires. If an individual is not made acquainted with onanism by comrades, books, or in any other manner, he will enter the age of puberty without any erotic thought, and is what is commonly termed "perfectly innocent." Notwithstanding this innocence, he may, in one moment, be overtaken by onanism. Sometimes violent erections cause him to touch the genitals with his hands, and thus he learns onanism involuntarily in some measure.

Certain movements, performed for some other purpose, may induce onanism. I have a vivid recollection of the history of a youth in whom I was quite specially interested, and who involuntarily committed an act of onanism. He was an industrious student, sixteen years old. During his studies he would take the most comfortable positions, and thus, lying one day on his stomach over three chairs, being lost in thoughts over his reading, he swayed his body to and fro, without noticing the erection. To his great surprise he experienced a most pleasant sensation in his genitals, quite new to him, but at the same time he felt a strange moisture. It was sperm, which he first took for urine. By such ways

---

[1] A. Theod. Stamm, Dr. med., Dr. phil., Die Verhütung der geschlechtlichen Ansteckung. Zürich, 1886, p. 63.

onanism may be induced in quite immature individuals, or individuals who have scarcely attained puberty. The incident at first is agreeable; its dangers are either unknown or the subjects are not willing to believe in any. Youths, however, who are well aware that the evil habit injures them, have seldom the courage to rid themselves of it, and finally the bad consequences make themselves felt.

We have now seen that the principal cause of this general spreading of onanism is innate in human nature itself, and is most commonly seduction. If, then, we speak of further causes of this disease, we shall understand thereby rather the occasions that help in the development of the germ. They do not cause onanism, properly speaking, but rather its immoderate practice.

The leader in these causes of onanism is the *sedentary mode of life* to which our youth is generally condemned, —*the want of out-door exercise*, the curse of the present customary mode of education.

Flogging on the bare back or buttocks is apt to incite to premature activity the sexual organs, or, rather, the nerves which lead from the center of erection through the spinal cord. This alone ought to be reason enough to induce us to abrogate as much as possible the brutal and absolutely unnecessary whipping of children. On some future occasion we shall show that such blows applied on the back and buttocks constitute a brutally empiric aphrodisiac.

Then, too, we have *lascivious reading and pictures* which, heating the imagination, for the time being incite the sexual instinct in the highest degree, and must be the cause of many excesses in onanism. Theaters, balls, and other entertainments where young men come in direct or indirect contact with the female sex are rather effective preventives against immoderate onanism, in spite of the demonstrations of all the misanthropic or thoughtless persons who never see anything beyond the sphere of their rooms.

I must raise my voice against those numerous authors who denounce riding on horseback as a cause of onanism. Riding, like every bodily exercise, is not conducive to onanism, and whoever asserts differently gives a proof of his want of knowledge on the subject. These gentlemen ape Lallemand always and in everything, while they also seize every opportunity for ridiculing his views.

As determining causes of excess in onanism I must denounce *idleness* and loitering about, which occasionally become habits even with very assiduous youths; also going to bed too early and arising too late, especially lying in bed awake in the morning. Again, *stimulating foods and drinks* may prove to be determining causes for excesses in onanism.

Another powerful cause in the propagation of onanism is the temporary or permanent impossibility of procuring a woman. The concupiscent desire increases, and finally even an energetic man of character may succumb. Again, *uncleanliness* may cause an accumulation of a sebaceous mass between the prepuce and the glans, inducing continual itching, and thus the hands will be induced to manipulate the member, this possibly ending in onanism.

As other causes of onanism may be mentioned stone in the bladder, various kinds of irritation about the promontory and the neck of the bladder, and certain *cutaneous diseases*, especially phimosis and balanitis. Oxyuris worms also, by setting up an itching or irritation about the genitals, may cause onanism. Since Lallemand, however, too much stress has been laid on the frequency of these causes of onanism.

Some authors blame the working of *sewing-machines* for causing onanism. The possibility of this cannot be denied, but, as Dr. Decaisne[1] says, a good deal of willingness is necessary.

In the last instance we shall mention too *tight garments* as a cause of onanism. These may, by means of friction, lead indirectly to onanism, but only if the genitals are already in a state of irritation.

In some exceedingly rare cases an inherited neuropsychopathic predisposition may, in very young children, cause premature awakening of the sexual instinct. The inevitable result will be onanism.

The **consequences** of onanism never tarry long in showing themselves. Since, in most cases, undeveloped individuals are given to excess in onanism, the whole organism soon suffers from the excitement and the continual loss of sperm. These stimulations are too frequent and too violent for the undeveloped nervous system. The change of substance in the body is rapid enough, so that recuperation from the frequent loss of substance would not cause much disturbance in the system if the individual were not in great need of all his force and all the energy of his metabolism for his growth and further development. Hence we see that delicate and sickly individuals suffer under excess in onanism a great deal more than strong and healthy ones.

First *anemia* is produced, under the influence of which the digestive activity suffers, and, consequently, the whole

---

[1] Fournier, De l'Onanisme. Paris, 1885, p. 67.

body. Again, the shock resulting from the practice of onanism enervates the whole system, and, less frequently, causes in the nervous system a state of excitability. The nervous system has but little power of resistance, and the shock is repeated very frequently.

Many forms of *neurasthenia* have their origin in onanism. With young children onanism may even cause far more profound disturbances of the nervous system. I have seen a boy nine years old suffering from epileptiform attacks arising from onanism only; the attacks ceased upon the application of an ingeniously arranged bandage. I doubt, however, that onanism can cause real and permanent epilepsy. The effect of onanism upon the nervous system is specially pernicious in cases where children begin onanism before there is any sperm. Here there can be no question of loss of substance. The time of production of sperm is in such cases considerably accelerated.

The assertion that onanism causes tabes dorsalis is certainly an extravagant fancy. To attempt to prove such an assertion by stating that of one hundred and nine cases of tabes, ninety-seven of the patients confessed to onanism, as Fournier[1] says, is a mistake. Fournier's report proves only that of one hundred and nine people at least ninety-seven have practised onanism at some time, and that most of them are ever ready to look upon any disorder in the conditions of health as a result of this wretched habit. The consequences of onanism are serious enough; there is no necessity to add anything from sheer fancy.

As to the special effect of excesses in onanism, we

---

[1] Op. cit., p. 125.

state, in the first place, that the *sensibility of the sexual organs is heightened*, and, in time, to an excessive degree. Very soon a hyperesthetic form of sexual neurasthenia develops, which, by itself, can cause impotence.

Edward Martin[1] says, "Masturbation unquestionably determines at first acute hyperesthesia and hyperemia of the prostatic urethra. This is ultimately followed by a chronic congestion and by almost complete anesthesia of the same region." I, for my part, never was able to observe chronic congestion and any degree of anesthesia together, but always found that, whenever there was congestion, hyperesthesia never was missing.

Besides, excess in onanism, in a much higher degree than excess in venery, is a cause of a gradually increasing *frigidity and aversion for the other sex*. This aversion is at the same time a cause and a consequence of impotence. The patient is impotent because he feels an aversion for the female sex, and he is averse to the female sex because he feels himself impotent. In the midst of the absurd and crazy social institutions of our time such people often assume the rôle of heroes of virtue, and are set as examples for others who have the misfortune (?) of being somewhat virile. Again, excesses in onanism are more frequently the cause of perverted sexual sensation than are excesses in venery.

The augmented sensibility of the sexual organs is the chief cause of the pollutions that invariably follow excesses in onanism; these pollutions are finally followed by spermatorrhea, and have impotence as their consequence. A relaxation of the entire sexual apparatus

---

[1] Impotence and Sterility. Hare, System of Practical Therapeutics, vol. iii. p. 662.

occurs simultaneously with the over-excitement of the sexual nerves and their centers. In this enfeeblement, as a matter of course, the muscular apparatus of the ductus ejaculatorii participates, whereby a further condition arises tending toward excessive pollutions and spermatorrhea. Of course, enfeeblement of the whole organism and weakening of the sexual organs go hand in hand and keep nearly apace with each other. Sometimes the rest of the body still presents a quite satisfactory state, whilst there is already great disturbance in the functions of the sexual organs. This is the case chiefly with onanists of mature age who are affected by hereditary low sexual power of little resistance.

Finally, it is asserted that excessive onanism may cause *atrophy of the testicles*. Such cases being described by Curling and Albert,[1] I will only point to their extraordinary scarcity, without giving expression to my doubts. It is more probable that atrophy of the testicles did not occur until onanism had caused impotence.

The most frequent and therefore best observed diseases caused by excess in onanism are immoderate **pollutions and** consequent **spermatorrhea.** Knowing, as we do, that onanism is an evil of long standing, we shall not be surprised if Moses knew of the pollutions.[2] According to his regulations, a man who had an effusion of semen was unclean for a whole day, and likewise the woman who may have been lying with him.

Too frequent pollutions are the most common causes

---

[1] Hofmann, Lehrbuch der gerichtlichen Medicin. Wien, 1881, p. 63.

[2] Trusen, Darstellung der biblischen Krankheiten. Posen, 1843, p. 9.

of impotence, and we may, therefore, say with some authority that onanism is the cause of impotence in the great majority of cases. Of one hundred persons suffering from pollutions there were at least ninety-nine addicted to onanism, so that all other causes together hardly ever come under consideration.

If a healthy, robust person, in full possession of his sexual power, does not in any wise satisfy his sexual wants, and if the glands preparing the sperm do not cease their action, an ever-increasing quantity of semen collects in them, causing a very great tension, particularly in the seminal vessels, and this leads to so-called physiological pollutions.

Now the question arises, How frequent may these be before they should be called excessive? Some authors have even pretended to fix the limits within which pollutions may be repeated. This is, however, not admissible, since the sexual organs do not act with the same energy with all persons, and the number of pollutions during a time of abstinence cannot be the same with individuals who were accustomed to daily intercourse, and with others who indulged only once a week. Therefore we have to consider as decisive not numbers but the circumstances that accompany and follow the involuntary loss of semen. Besides, the great divergence in the frequency considered normal by different authors is sufficient proof of their unreliability.

A pollution must be called normal under the following conditions: it must, first of all, occur during sleep,—*i.e.*, during absence of consciousness and will-power; it must be accompanied by a vigorous erection, erotic dreams, and by the natural sensual gratification; it must cause a sensation of well feeling and relief, but not of faintness, depression, headache, or other similar troubles. If any one of these conditions is missing, the pollution must be considered morbid.

We shall, with Curschmann,[1] classify morbid pollutions as follows:

I. *Morbid pollutions occurring during sleep.*

a. The pollutions are *more frequent* than is normal according to the peculiarities of the individual and the natural state of his semen secretion. The accompanying phenomena are unchanged, but the patient feels afterward faint, low-spirited, and is sometimes troubled with headache, etc.

b. The *number* of pollutions reaches such a height that they appear for this reason alone as pathological. The pollutions may occur every night, or even more than once in one night. Moreover, they may occur sometimes directly after coitus, and even in a bed that is shared with a woman. The accompanying phenomena still resemble those of normal pollutions, but the consequent pathological sensations are still more marked than in group *a*.

c. There is very great frequency, but an *absence of the phenomena accompanying normal pollutions*, such as erection, erotic dreams, and voluptuous sensations. The ejaculated semen is very small in quantity, and in quality a thin liquid. In this group of morbid pollutions the psychical shock and the loss of substance are both insignificant, and yet the consequent phenomena are very grave.

---

[1] Die functionellen Störungen der männlichen Genitalien. Ziemssen's Handbuch der speciellen Pathologie und Therapie. Band ix., Hälfte 2, p. 467.

II. *Morbid pollutions in the waking state.*

a. The pollutions take place while the individual is awake, *in consequence of trifling mechanical irritation*, such as friction by a tight garment, riding on horseback or in some conveyance.

b. The so-called diurnal pollutions happen even under the impress of *psychical influence.* Finally, as the last form:

c. The patient loses semen during *micturition or defecation.*

In reference to the above and similar classifications it must be remarked, however, that loss of semen during the waking state occurs frequently without coming under the head of an aggravated state of morbid pollutions. Spermatorrhea, particularly following an obstinate gonorrhea, is to be considered a case of this kind.

Most authors distinguish between pollutions and spermatorrhea, but their definitions of spermatorrhea are quite varying, the limit between pollution and spermatorrhea being fixed differently by different authors. Since, however, these distinctions are of no value in therapeutics, we shall, with Curschmann, give the name of spermatorrhea to those forms of pollutions that are of a high degree and take place during consciousness. Roubaud[1] gives the name pollutions to those losses of semen which are accompanied by venereal orgasm, and spermatorrhea to those that are unaccompanied by sexual desire or erection or voluptuous sensation. Older authors have described by the name spermatorrhea a state in which sperm is said to flow continuously. Since we have no trustworthy report of any such case,

---

[1] Op. cit., p. 327.

we cannot have any faith in the above statement, although we admit the possibility that, if the organs of ejaculation have ceased to act, the sperm may flow off as it is secreted. We must lay special stress on the fact that although morbid pollutions occur very frequently, yet those of a high degree which deserve the name of spermatorrhea are exceedingly rare.

To Lallemand is due the merit of having directed the attention of the medical profession to this disease of morbid pollutions, which till then had been well-nigh neglected, and we can gladly overlook the exaggerations to which he may easily have been misled. Modern authors profit by his investigations, and yet ridicule the man who has written his name indelibly on modern pathology and therapeutics of pathological sperm effusion. This keen observer has studied and described the nature of this disease so precisely that the authors of to-day can only express in other terms what has already been said by Lallemand. If they deviate essentially from his path, they are generally in the wrong. His method of cauterizing is still the alpha and omega of the treatment of this disease, although variously modified and named. It would be an injustice if we were to blame him for seeing in every one of his patients an individual affected with spermatorrhea, and for having an unlimited confidence in his method of cauterizing. Some of our celebrated specialists have similar weaknesses.

Lallemand unquestionably deserves the credit of having proved that excessive loss of sperm is a disease or, at least, a symptom of disease, which can and must be treated. Of what use to a man who suffers from weakening pollutions is a physician who, following the example of renowned clinicians, laughs at him and sends him home with some insignificant and useless prescription? You should see the despair that takes hold of such an unfortunate being when he sees what little importance the doctor attaches to his condition, which, he feels, is sapping all his physical and mental strength.

It is a poor consolation to say that only the minority of the patients asking for advice about losses of sperm are really suffering from pathological pollutions. Such assertions coming from competent men [1] are apt to make the physician careless or indifferent in the treatment of this so eminently serious disease, and to induce him to consider these pitiable persons as sufferers through the imagination. I am, on the contrary, of the opinion that such patients, with but few exceptions, are in great need of medical help for these excessive pollutions, which may not kill them at once, but which cause an irreparable loss of general and sexual force. The sooner such patients consult the physician, the more wisely they act, because chronic evils are more difficult to treat.

Eulenburg's [2] opinion seems to me very plausible. He says, "I believe that a pollution is no more to be considered as normal than a cough or vomiting, and that even the so-called normal pollutions originate really in some unusual and exceptional irritation, which may be comparatively light, but which acts upon the center of ejaculation."

It has already been pointed out that onanism is nearly the only cause of morbid pollutions. Cases, however, may be met with in which morbid pollutions are also caused by acute inflammatory conditions in the urethra, such as

---

[1] Curschmann, op. cit., p. 495.
[2] Sexuale Neuropathie. Leipzig, 1895, p. 55.

gonorrhea, inflammation, and tumors of the seminal vesicles, chronic inflammation of the neck of the bladder and of the pars prostatica urethræ. Phimosis, various diseases of the rectum, such as piles, fissures, eczema, and other cutaneous eruptions of the rectum and its vicinity, are, since Lallemand's time, accused of being the cause of morbid pollutions. Such causes as these are seldom observed and taken into account, although the accumulation of smegma in consequence of phimosis, the presence of oxyuris vermicularis in the rectum, or other diseases producing irritation in or about the genitals, may easily cause erections and also nocturnal emissions. We have already seen that such diseases are causes of onanism, and, indirectly, of pollutions.

It is a disputed question whether *obstipation and difficult defecation* are capable of causing pollutions and spermatorrhea. It seems very plausible that the pressure of hard and voluminous scybala and forced contraction of the rectum may force out some seminal liquid. There is no doubt that masses of feces which have collected in the rectum can cause erections and nocturnal emissions in consequence of the pressure and incitement they may exercise on the sexual organs. It is more than doubtful, however, that with a healthy man the difficulty in connection with defecation is able to press sperm out of the seminal vessels. We would rather agree with Curschmann's[1] view, which he expresses thus: "Theoretically considered, the opinion made reference to— viz., pressure from the rectum on the seminal vesicles— is not so plausible as it would at first appear. These vesicles are so placed between the bladder and the rec-

---

[1] Op. cit., p. 488.

tum that they have free motion, and can thus easily give way to pressure coming from the rectum, so that the latter would be more likely to exercise any action on the closely joined and well-fixed ducts and their orifices than on the widely diverging blind ends of these formations. But pressure on the orifices of the seminal ducts would rather have a closing effect than otherwise."

The explanation of the origin of this liquid would have far more anatomical probability if applied to the prostatic humor. The prostata, as is known, is firmly fixed in the pelvis, and so placed between the fundus of the bladder and the expansion of the rectum, situated immediately above the anus (and particularly well developed towards the front), that the hard feces must almost of necessity be pressed against the prostata by the pressure of the sphincter ani, which acts from before backward. I had repeated opportunities of examining microscopically sperm-like liquid which different men emitted during difficult defecation, and which they believed to be semen. In but few cases could I find any seminal filaments, and in these cases I also discovered other evidences of spermatorrhea.

Very seldom are morbid pollutions produced by *general diseases*, as anemia, general debility, and neurasthenia. By the side of all these causes, which rarely come under observation, onanism remains almost the only cause of morbid pollutions that invites closer medical attention.

Why does onanism cause pollutions? We have already answered this question. The conditions that make onanism more likely to cause morbid pollutions are the youthfulness of most onanists and the facility with which they can, at any time and in any place, indulge in their evil habit. Excessive onanism, or onanism in general with individuals whose power of resistance is still low, causes, almost without exception, a state of slight inflammation about the ductus ejaculatorii. This inflammation produces in the vessels that convey the sperm an irritability of so high a degree that it is altogether out of proportion to its cause, and this irritability is, with rare exceptions, the cause of the morbid pollutions. The congestion probably never reaches such a point that a catarrhal secretion could be ascertained.[1] A slight swelling and reddening of the pars prostatica urethræ can always be traced, and upon this fact we may establish, with the greatest plausibility, our diagnosis of excessive onanism, if we exclude any previous gonorrheal inflammation of the urethra. It must, however, be remarked that the pollutions, if they last a long time, cause by themselves chronic inflammation of the caput gallinaginis. This swelling and reddening will generally disappear as soon as the morbid pollutions begin to pass over into that stadium which is commonly called spermatorrhea. In this the caput gallinaginis turns pale and atrophies. This same cause probably induces the relaxation of the orbicularis muscles of the ductus ejaculatorii, which is followed by a dilatation of the same, and, finally, by spermatorrhea.

These local changes are probably the immediate cause of the pollutions, but neither can, with any reason, be considered an independent disease, as they are merely symptoms of the disease consequent on onanism.

We have seen that the pollutions are caused by the

---

[1] Fürbringer, Eulenburg's Real-Encyclopädie. Wien und Leipzig, 1888, Band xiv. p. 596.

irritation as well as the *relaxation of the spermatic passages*. An attempt has accordingly been made to distinguish two forms of morbid pollutions. However valuable it may be in therapeutics, this distinction cannot be accepted as valid, but must give way to the assertion that the states of irritation and of relaxation are different stages of one and the same disease. The former always precedes the latter; but both occasionally exist at the same time.

When once the pollutions have become chronic, then we have also to deal with the *influence of habit*, which is sometimes quite incomprehensible in the case of many organs, and offers an obstacle to the cure of pollutions even when their causes have been removed. There are individuals who are *naturally predisposed* to morbid losses or effusions of semen, in whom they are caused by the most trifling excesses.

The *diagnosis* of morbid pollutions is an easy task, but the diagnosis of spermatorrhea is frequently more difficult. It is absolutely indispensable that the semen be examined microscopically, whether discharged by ejaculation or otherwise, and the patient himself must be subjected to an examination with the endoscope.

The results of such sperm examinations from morbid pollutions are very various. Spermatozoids are found in variable quantity, being absent only in spermatorrhea of a high degree. In the same patient suffering from morbid pollutions there are found, at one time, quite well-developed spermatozoa, appearing dark in the field of vision and provided with long tails, while at another time almost nothing but young formations with light, water-colored heads and short tails. A remarkable fact is that after a pause of several days between pollutions you will find the seminal filaments very sparing in the ejaculated semen, whilst the filaments appear more numerous in sperm from pollutions repeated after short intervals.

I have made numerous microscopic examinations of sperm coming from pollutions. It has been mostly sperm from prisoners, and only in a few cases from private patients. These examinations have convinced me that the spermatic fluid of morbid pollutions does not differ from semen ejaculated otherwise, in the quantity or in the form of spermatozoa, if we exclude the comparatively rare cases of spermatorrhea of a high degree. The difference is only in the lesser vitality of these filaments.

The filaments in sperm ejaculated during intercourse are partly alive after a lapse of forty-eight hours, if the fluid has been preserved under favorable circumstances; those from pollutions are, without exception, dead after a few hours. Even spermatozoa coming from pollutions which the majority of authors would call physiological have much less vitality than others discharged in coitus, for instance. This can easily be demonstrated by the examination of semen from the same person, but discharged at one time in pollution and at another in coition. My experiences in this line are not in conformity with Fürbringer's[1] views on the relations of the products of the prostata to the seminal fluid, though recent and at this time unfinished experiments have caused some doubts in my mind.

The result of *endoscopic examinations* has been spoken of already, and we here add merely that, since the

---

[1] Die Störungen der Geschlechtsfunctionen des Mannes. Wien, 1895, pp. 8, 9.

urethra of a patient affected with spermatorrhea and pollutions is very highly sensitive, the introduction of a sound or of an endoscope necessitates the greatest care and gentleness possible.

The result of an endoscopic examination is negative only when the morbid pollutions are caused by hyperesthesia of the sexual centra,—a case which is certainly very rare, since an individual whose ejaculatory organs are in good condition may lie the whole night with an erection and be visited by erotic dreams without suffering any loss of sperm. The result of such examination with an endoscope may also be negative if the morbid pollutions have appeared among the sequelæ of some nervous disease; but this also would be true in the beginning only, because, if the pollutions continue during a longer period, changes about the ductus ejaculatorii begin to make themselves noticed. On the whole, the cause of pollutions has very seldom to be looked for in the direction of the centra, because, in the great majority of cases, the excessive losses of semen are occasioned by local disease.

In most cases of morbid pollutions the *exterior appearance of the sexual organs* points to morbid changes. The penis and testicles with their surroundings have generally a flabby, withered look; the testicles hang lower than they should, and are sometimes sensitive to even light pressure. Almost without exception there is a diminution of the warmth, sensitiveness, and irritability of the exterior sexual parts.

The *general state of health* and appearance of the patient are more or less sickly in proportion to the grade of disease. Here and there you may see persons having a very healthful appearance who are nevertheless affected with morbid pollutions of a high degree. When you set your eyes on such a miserable, pitiful being, who, although gifted with some power of resistance, nevertheless ends by losing health and sexual power, you involuntarily ask yourself whether he can possibly be the object of such poor raillery as may be read in Niemayer's "Manual of Special Pathology and Therapeutics."

The patient is sickly in appearance, and presents the picture of exhaustion in most cases,—without exception in cases of a high degree. The chief disorders or troubles make themselves felt in the digestive organs and in the nervous system, the deleterious influence on virility being constant and finally destroying it totally.

In this diminution of virility and in the incessant pondering over the loss of semen, repeating itself without remission and driving the patient to despair, is most often to be found the cause of the changes in the character which are nearly always observable in such patients.

.Although a rational investigator cannot, even by a thorough study of the pollutions, discover any advantage in this kind of discharge of sperm, yet there have been amongst authors ones who, while arousing a horror of every copulation outside of wedlock, have gone so far as to speak of the pollutions as a wise provision of nature.

Thus there arise the questions, Are the pollutions necessary, and Are they of any advantage to man? "No" is the answer to both questions. I will not speak of the wasteful spilling of the precious fluid which had better be used for the creation of new human beings, since it would be a real misfortune for mankind if all sperm were so used. The waste of sperm has its cause in the course of nature; but another circumstance claims our attention. The sexual power of man is one of those few real pleasures of our existence, and the pollutions deprive many a one of a considerable share of these enjoyments, which are in any case only sparingly meted out. Pollutions should never be allowed to exist in any one, for they can have only one of two things as their cause : either they prove a real want that is not satisfied, or they are the symptom of some disease. In the former case the individual concerned should do his utmost to obtain his share in the enjoyments of life ; in the latter case the sufferer ought to seek help, as it is his duty to do. The main cause of the lack of spirit and the helplessness of these sufferers lies in the present views of the leading professional spirits. If they show any feeling for these pitiable individuals, it is a sympathizing shrug of the shoulders at best, or it may be a scorn, since life is not at stake.

My conviction is that pollutions will soon be cancelled from the list of physiological functions, and treated as *a pathological symptom*. I am so much the more inclined to this view, as I am not aware that pollutions have ever been noticed in animals.

The injurious effect of sexual excesses on the functional capacity of the sexual organs is an admitted fact, and has been generally rather exaggerated than underestimated, as is the case to-day. Many men of the medical profession, who felt it incumbent to play the rôle of moralist, have at all times decreed the most horrifying chastisements on disobedience to the sixth commandment of God. These gentlemen of such high morality have too often encroached upon the legislative power, but, luckily, they were not invested with executive power, and hence it comes that many members of frail humanity, so sinful ever since the time of Adam, and

Eve, continue unpunished, most of them meeting with punishments far less severe than those ordained in the penal codes of the above-mentioned legislators.

On all sides and at all times has it been emphasized that nature resents every infraction of her laws. This is essentially correct, but it is remarkable that most of the authors look upon such infraction as equivalent to excess only, and few of them state that not excessive indulgence alone but also excessive **continence** can harm the body and the sexual power. An explanation of this omission is chiefly to be found in the fact that real continence is practised so seldom. It is well enough to exalt absolute continence, but its great rarity makes talking of it a positive waste of time. Thus I have no great faith in absolute continence, and believe the continent, with very few exceptions, to be onanists. Accordingly, I do not wish to speak of absolute, but only of relative, continence.

In theory it will be easy to understand that sexual faculties not kept in sufficient practice are weakened thereby, and this for several reasons. Every gland, and consequently also the sexual glands, requires a certain amount of excitation of its nerves in order to produce energetic action. Every muscle, and consequently also the muscles of erection, can become strengthened only by exercise. All bodily functions demand appropriate gymnastics, the sexual functions no less than any other. It is quite noteworthy that authors even in our time are rather loath to advance these truths, and they forget in some measure that " continued inactivity of nerves diminishes their irritability even to annihilation." [1]

---

[1] Landois, Lehrbuch der Physiologie des Menschen. Wien und Leipzig, 1893, p. 674.

FORMS OF IMPOTENCE.

A robust man with well-developed virility and powerful sexual instincts will never be in danger of making too sparing a use of his procreative power, at least not voluntarily. Such is more likely to occur with people who are originally poorly provided with sexual strength and desires; and these are the very ones who cannot afford to do without reasonable gymnastics in sexualibus, just as weakly children require bodily gymnastics more than stronger ones do. Lallemand[1] expresses this very appropriately when he says, "No one will think that a delicate child ought to be kept from gymnastics because it shows in comparison with its comrades less inclination and aptitude for all kinds of bodily exercises."

Since the sexual power plays such an important rôle in human life, it would seem natural that some efforts should be made to strengthen it. We are yet, however, at a great distance from practical and unprejudiced views.

It is practically proven that continence, whether absolute or relative, induces a weakening of virility. This fact, which is in accord with theory and proven by practice, may seem to contradict the very common experience that powers injured by sexual excesses recover during continence. This contradiction, however, is merely apparent, because in impotence resulting from sexual excesses the cause is not in the diminished secretion of the sexual glands or in a lowered capacity of the muscles, but in the temporarily weakened state of the nerves. During moderate continence the over-strained nerves and their centers have the necessary time to become calmed and strengthened; the glands and muscles

---

[1] Pertes séminales, tome ii. p. 255.

cannot be injured in their capacity for action during a time of rest that is of only short duration. Continence is certainly not of equal importance with sexual excesses, not because it is less pernicious, but because of its greater rarity.

The commonest consequence of absolute or relative continence is weakening of virility. Sometimes this weakening is preceded by a stage of great irritability of the sexual organs, during which too frequent pollutions may set in and become permanent. Thus nature helps itself, but, of course, not without injuring the organism in another direction or way, since pollutions are never unattended by pernicious consequences. In general, the sexual instinct disappears gradually if not roused from without.

Absolute continence is so seldom the object of medical observation that we cannot say anything definite about the phenomena accompanying it. In this we also find the only explanation for the song that Acton sings in praise of continence. As to those who are approximately continent, daily experience informs us that they are seldom endowed with marked virile power, and I believe that they are naturally possessed of a low degree of sexual power, because a duly gifted man neither will nor can be continent. Weakness and incapacity are sometimes arrayed in the garment of virtue. "In any case there is a close connection between the activity of the generative glands in a pubescent individual and the degree of his libido."[1] Nowadays there are probably few who still believe the obsolete fable that sperm once secreted can be reabsorbed and then be of special benefit to the body.

---

[1] Krafft-Ebing, Psychopathia sexualis. Stuttgart, 1886, p. 30.

Elsewhere we have already stated, as far as our knowledge extends, what the internal processes during continence are, and what becomes of the sperm stored up in the glands that prepare it.

The cases that are oftenest observed and that afford the clearest proofs of the weakening influence that continence exercises upon virility are those in which robust men are compelled to observe continence. In this respect I have been particularly favored by having the opportunity of making my observations during the partial mobilization of the Austrian army when a part of it was for a time stationed in Bosnia. Nearly all the officers, friends of mine, vigorous young men, told me that at first it was hard to submit to the abstinence forced upon them by the social circumstances; but after a time it was comparatively easy to bear. Even the young gentlemen were not surprised at their ability to abstain, but their astonishment followed soon when one or another obtained leave of absence and expected to do wonders when at home. Instead, he had rather to remain on the defensive at first, at least, until the novel excitations had again animated his sexual organs to new activity.

Edward Martin[1] mentions prolonged continence as one of the causes of atonic impotence in cases where the instinct is strong and where the mind has long been given up to amorous desires. This author thinks the reason for it lies in the "prolonged congestion which does not receive its normal physiological relief." The same author states further that "in some cases the organ is so poorly developed that a successful intercourse is

---

[1] Impotence and Sterility. Hare, System of Practical Therapeutics. Philadelphia, 1882, vol. iii. pp. 661–663.

well-nigh impossible. This is generally observed in those who have been continent. In such cases local exercise may act as beneficially as it does upon other parts of the body."

Having now discussed the more frequent causes of neurasthenic impotence, we shall treat separately a few of the more prominent *forms*.

All kinds of sexual excess lead frequently to the different grades of *paralytic impotence*. Complete paralysis of the sexual nerves and centra occurs probably rather seldom as long as the rest of the body keeps healthy and robust. In senile impotence it is constant. More frequently we notice the other numerous forms of sexual neurasthenia, every one of which may pass over into paralytic impotence.

In the first instance we have the so-called **irritable weakness**, which is on the confines between diurnal pollutions and sexual neurasthenia, but differs from morbid pollutions chiefly by the absence of any material change in the sexual organs. The irritable weakness consists in a sexual irritation which is generally of a high degree, but with which erection does not keep pace, for erection is either incomplete to start with or it becomes complete only after long exertions. In either case, however, the ejaculation is precipitate and occurs even before the introduction of the penis into the vagina, in case the disease is high-graded. This disease must not be confused with a precipitate ejaculation with incomplete erection that may occur quite normally with most men after an unusual abstinence.

We need not add anything about the disagreeableness of this disease, but it is the clearest proof of the incorrectness of the assertion that the pleasure lies in the ejaculation alone, which opinion has been shared and upheld by many authors. Premature or precipitate ejaculation deprives the man, and still more the woman, of the due pleasure; it may, moreover, be an obstacle to conception, because with the few motions of coition, or even their entire absence, there is no orgasm caused in the woman, this being generally a condition essential to conception.

If the precipitate ejaculation is caused by disease of the ductus ejaculatorii or of the colliculus seminalis, it remains constant and repeats itself without exception at every coition. If the precipitate ejaculation is based upon irritable weakness as a form of sexual neurasthenia, then it is varying like all other neurasthenic diseases. The ejaculation may then take place, at one time before the penis enters the vagina; at another time it may occur after a few movements; again, the coitus may be accomplished quite normally, and at some other time ejaculation may even be delayed.

In purely neurasthenic cases of irritable weakness we shall not find that high-graded erethism of the urethra which Ultzmann[1] speaks of, and which can always be noticed when the "irritable weakness" is caused by material changes about the colliculus seminalis.

There are neurasthenic individuals who have precipitate ejaculations with one woman, though they can have quite normal coitions with other women. Most neurasthenic persons with difficulty get over the first attempts at coition with a new acquaintance; these are generally unsuccessful attempts. Such men must first get ac-

---

[1] Potentia generundi und Potentia cæundi. Wiener Klinik, 1885, Heft 1, p. 25.

customed to their new acquaintance. Their very vivid imagination must first be somewhat pacified; then they recover their normal condition for the time they keep company with that same woman. This, however, is not lasting with most neurasthenic patients, because they soon conceive an aversion for the woman in question, and their disease is thus the cause of their inconstancy.

The irritable weakness attended by material changes in the sexual apparatus should, in my opinion, be classed rather with morbid pollutions. It has its cause nearly always in excessive onanism. Such individuals, besides, suffer almost without exception from frequent pollutions.

Excessive onanism, pollutions, sometimes onanism habitual though not excessive, are in like manner capable of causing neurasthenic irritable weakness. It is not onanism exclusively that causes this form of disease. Sometimes a congenital predisposition to neurasthenia, or other neurasthenic disorders, cause temporary or permanent irritable weakness. At other times the cause may be mental onanism and excitement immediately preceding coition. There are, besides, single forms of irritable weakness that originate in or arise from an over-sensitiveness of the glans; this is the case especially with an individual whose glans is wholly covered by a prepuce which can be retracted with moderate effort, and the glans is exceedingly sensitive when it has thus been laid bare.

When considering and estimating such conditions, it must not be forgotten that whenever there are signs of increased irritability of the nerves, we are nearly always in the presence of the first stages of deterioration of nerve energy. From all the above circumstances we must infer that it is not true that all forms of impotence through irritable weakness are "spinopherous or preponderantly spinal."[1]

A rather common form of impotence is that kind of sexual neurasthenia which is generally called **psychical impotence,** which might with more propriety be named *hypochondriac impotence.* We very seldom see a purely psychical impotence in which the sexual organs and their connection with the central organ, as well as the entire nervous system, are perfectly sound. There may be very impressionable individuals who are apparently healthy in every respect, and yet may become temporarily impotent simply by the effect of the thought that they are impotent, or by the fear of not being able to give satisfaction in a certain case. Such individuals are evidently healthy in appearance only; but, like a hypochondriac who suffers from an evil that is real, though he magnifies it, so also is the psychically impotent man afflicted with some organic defect which is not discovered until he is examined with care. Then it is found without exception that these are neurasthenics suffering from psychical impotence, and are generally people who have weakened their sexual organs and burdened their conscience through onanism or some other mismanagement of their sexual power. They need intense excitement to obtain an erection necessary for coitus. Their centers of erection must be blunted and their inhibitory centers just as sensitive, since a mere thought is often sufficient to excite the latter and paralyze the former.

In psychical impotence it must further be remembered that every thought likely to divert the mind from the act in contemplation can also prevent erection altogether or

---

[1] Eulenburg, Sexuale Neuropathie. Leipzig, 1895, p. 28.

overcome it if already begun. Hence it is a matter of course that anxiety, fear, shame, or any other feeling that may engage the mind cannot be favorable to erection; and the thought alone that one may not be able to accomplish coitus can become a hindrance to erection. Neurasthenic patients, also described under the name of hypochondriacs, are not very capable in matters of sexual functions on account of their ever-present ill humor and anxiety.

It would be a great error if we were to assume that only weakly and sickly persons can be affected with psychical impotence. We see often enough quite robust and vigorous men subject to occasional attacks of psychical impotence. It renders them exceedingly miserable; they cannot get rid of the idea that they are impotent, and in the end positively become so, if some incident or sensible advice does not convince them to the contrary.

The phenomena of psychical impotence are manifold and variable. We could not expect that it would be otherwise in any neurasthenic disease. At one time the patient may have vigorous erections, but they come at the wrong time; then, when the occasion presents itself, there arrives instead of the erection the thought of the impossibility of the accomplishment of coitus. At another time the erection is there when wanted, but fades gradually as the patient proceeds to the act, or at the first movements. Quite remarkable is the circumstance that this form of impotence may also affect quite strong and vigorous individuals, but only on the occasion of new sexual connections.

There are many men, and I am inclined to include here all *bon-vivants*, who experience great difficulty in accomplishing a first coitus with any woman. It is remarkable that in this disease—for a disease it most certainly is—a man may feel a great preference for a woman, wish most ardently sexual union with her, and yet remain impotent when in her arms; when, on the other hand, he may, on leaving her, go straight to an undesirable woman who inspires no love, and accomplish the act without the least difficulty, and even repeat it. The only reason is that in the latter case old acquaintance makes him feel at home and excludes any apprehension of mishap. The cause cannot be a sense of too intense love or reverence, *bon-vivants* not often being influenced by such feelings.

The want of responsiveness in a new acquisition has sometimes a disadvantageous effect on the man; nearly always it is the apprehension of a failure that makes men who are generally expert impotent for the time being. To this is added, in illegitimate sexual connections, the chance of unfavorable circumstances, such as want of time, rest, convenience, and responsiveness. These men generally recover their former vigor perfectly after having once succeeded with a new acquaintance, and they then console themselves with that proverbial German saying that nearly always proves true, "Aller Anfang ist schwer."[1] If a man has cohabited for a long time with one woman, the beginning with another is attended with some difficulty; hence the trouble experienced in the first matrimonial infidelity, which often drives a husband back to the arms of his better half.

Inexperienced young people may in the beginning of matrimony remain psychically impotent for some time. This inexperience must, however, be of a high degree, if

---

[1] "The difficulty is in the outset."

a mixed form of impotence is not at the bottom of the trouble. Cases of impotence caused merely by inexperience are no doubt very rare, whatever books may have to say on the subject. Such people are generally young onanists, or they may have grown old in intercourse with public women without having an idea of proper coitus.

In this great variety of phenomena of neurasthenic diseases it will not be surprising if still other forms than the ones mentioned present themselves for observation. It would lead us too far if we should attempt to enumerate here all the phenomena of sexual neurasthenia. There are so many that one life is too short for the study or observation of them all. You may fancy you have exhausted the subject at one time, and yet new varieties crop up: there is no end of them.

We shall here describe but two more groups of sexual neurasthenia. These are generally designated as temporary and relative impotence.

By the expression **temporary impotence** is meant a condition in consequence of which a man may within short intervals of time be now virile and now impotent. The individual is utterly incapable of understanding how he became impotent, as only a few days ago he had all the necessary virile power. He now searches after every and any possible cause, and accuses liquors, tobacco, physical and mental exertions, etc. Of course, we leave out of question here that temporary impotence which visits particular individuals after the commission of sexual excesses, as it is in the nature of things that the energy of the sexual impulse is in an even proportion to the provision of spermatic fluid on one hand, and with the stored-up energy and elasticity of the nerves on the other. The proportion is even up to a certain degree; beyond that it becomes reversed.

With man the sexual instinct luckily does not make its appearance periodically, as it does with animals. Most authors assert that in spring it is most intense, and prove this assertion by lists of births with their dates.

Persons who are affected by intermittent impotence are generally sexually excited in a high degree, and most so when they are in a stage of impotence. This sexual excitement is not a real orgasmus venereus, always leading to an erection; but rather a state of agitation caused by mental onanism, a debauchery of the thoughts, as Lallemand calls it, "Licentiousness in thought which is in marked contrast to the impotency of the executive organs." The more agitated such a patient is, the more his penis shrinks, the less he is capable of copulation. Then, all of a sudden, when the sufferer has resigned himself to inaction, erection appears and coition may be performed, and sometimes even repeated. This form of transitory impotence may also come to quite healthy individuals during great agitation.

Relative impotence designates in general a condition in which an individual can accomplish coition only with one or certain mates, and is completely impotent with others; or a condition in which an individual cannot accomplish coition with one or certain women, and is perfectly virile with others.

Strictly speaking, there are few persons who are not relatively impotent, for only few can accomplish the act with all persons. In the same way the stomach may revolt against certain things; so may the nose. The specific odor of one person may be absolutely objectionable to another. The difference in the chemical elaboration in different individuals imparts a certain odor to the whole body, and also to separate parts of it. This may often be the cause of sympathy or antipathy between persons of different sex. This state is to be called morbid only when a man is incapable of accomplishing coition with a certain woman, although he is desirous of doing so, and has a liking for her.

We read in older works on impotence—and many modern ones have copied from them—various miraculous stories of how excessive love, a state of rapture of a high degree, has rendered the poor lover impotent. French authors particularly exhibit a very poetical conception of this state.[1] Such stories cannot, I suppose, be relegated altogether to the realm of myth, but occur, no doubt, very seldom, such a state of agitation, not at all in accord with our modern way of feeling, being generally of so short duration that no assistance of the medical art is required.

The case is very different with regard to feelings that are the opposite to those of love, as aversion or even hate. These are not of a nature to inspire a man with amorous lusts by the side of the object of abhorrence. We are more particularly interested in cases where these extreme feelings are not active.

There are many husbands who are impotent out of wedlock. With most of these gentlemen virility has not been duly developed to begin with. With many it is a kind of depravity and habitual pampering that has rendered them thus, and they would be perfectly virile out of wedlock, if time and circumstances in regard to other women were analogous. It is a similar sort of depravity

---

[1] Roubaud, op. cit., p. 377.

if a man can accomplish coition only in certain positions or after certain preparations. Among old bachelors, who pluck forbidden fruit only, there are some who must always have their lady-loves in full dress.

After all, there is certainly no harm in a husband's impotence outside of wedlock. It is rather preferable for both himself and his better half; but it is a very different case when a husband is impotent with his legitimate wife. This, Ultzmann[1] asserts, is a very disagreeable affair. Such relative impotence may exist from the very beginning of matrimony, or it may arise in the course of time. At the beginning of matrimony it never occurs in what we commonly call love-marriages, and only seldom in marriages of convenience. Occasionally the personal odor of a woman may be disagreeable to a man, "le parfum d'une femme" (Galopin). This may go so far as to prevent erection. In such a case, sound reason would demand neither more nor less than a divorce.

In the course of matrimony psychical and physical *differences* arise between husband and wife, especially in cases where the wife ages before the husband, these differences often inducing a relative impotence. Sometimes the cause of such relative impotence is in the circumstance that the husband has entered upon extra-matrimonial relations, which seem to him preferable, so that his enjoyment at home diminishes. All these are cases where the physician may have advice to offer, but no help.

There are cases, however, where the wife may be wanting in her duties toward the husband. She may not be sufficiently responsive, or she may have a repug-

---

[1] Op. cit., p. 25.

nance for sexual intercourse. Enjoyments not shared are in sexualibus less than half-pleasures. Thus, the husband feels less and less desire to repeat fruitless efforts, and looks elsewhere for indemnification.

PROFESSIONAL IMPOTENCE.

"Desidiam puer ille sequi solet; odit agentes."—OVID.

That vocation may have great influence on the sexual power of a man was known to the ancient hygienic legislators, as Moses, Mohammed, etc. In this respect certain vocations stand in bad renown, and especially those which require principally mental exertion.

In general we find the notion widely spread that the higher an individual stands intellectually, the less will be his capacity for sexual functions. La Fontaine's "un muletier à ce jeu vaut trois rois" has passed from the French books into the Italian. In my opinion this almost proverbial saying is not quite in accord with the truth. I admit that for a short time the victory may remain with the "muletier," but after a certain time the brutish instincts of the "muletier" will be appeased, his fantasy will not be able to take the place of the real wants, and then the "roi" will undoubtedly be the better man.

There is no doubt that in sexualibus the intellectual man stands above the peasant, for instance. Any one who has an opportunity to watch these country people in their hymeneal life will soon discover that, although the peasant may perform extraordinary feats at first during his matrimonial life, he, nevertheless, later on grows very neglectful in his conjugal duties. Of course, the hard work brings here physical fatigue, and there is little imagination at work. Again, the attractiveness of these honest countrywomen diminishes rapidly.

Heavy bodily work bringing about bodily fatigue, and intense mental exertion causing mental fatigue, are not favorable for the sexual power of man. Entire inactivity would be favorable, it is true, if we mortal beings did not grow weary in the *dolce il far niente*, which soon leads to devious roads and to excesses which then undermine sexual vigor. Those occupations in which labor is performed without fatigue and accompanied by a certain amount of mental exercise, and those mental occupations that are relieved by the requisite amount of physical exercise, are the callings that present the most vigorous men.

The Talmud treated the members of the learned professions to an exceptional favor by granting them the privilege of intercourse with their wives once in two or three years, while the wives of others had a legitimate claim upon their husbands at least once a week. Bookworms are generally weak in sexualibus, as has been known for thousands of years. Many a one may, like Rousseau, have received from some experienced lady the confidential advice, "lascia le donne, e studia la matamatica," without making so much ado about it. Part of the cause is in the sedentary life of these men, which is not conducive to health; and part is in the over-exertion of the brain, that is apt in the course of time to aggravate the neurasthenia which exists often *a priori* with men who are inclined to mental efforts. This, then, will in the end bring about a new diminution of the sexual power. Besides, intense mental activity exercises the thinking parts of the central nervous system to the detriment of the sensitive and motor parts.

If *artists and scholars*, as painters, actors, authors,

physicians, professors, etc., are sometimes reported to be not disinclined to love, or to be sensual, this is no proof of the contrary, because the presence of sensuality does not imply that of sexual strength.

Withal, we see that highly intellectual and learned men are often possessed of a considerable wealth of children, so that Nordau's[1] dictum, "From common men we obtain the conservation, from great minds the intellectual advancement, of our species. The same individual cannot be equally capable of producing both thought and children," merely repeats a popular dictum not supported by fact. The best thought has come from people who have also produced children; while but few impotent men can console themselves with the fancy that their weakness has not prevented them from producing at least good thought.

Some authors advance the idea that people who attach great importance to an excellent table, the so-called *epicures*, possess but little sexual impulse. For my part I must state that I have observed the opposite condition only,—the so-called great eaters are also great worshippers of the fair sex until the time when the superfluous quantity of nourishment induces obesity, and obesity grows from day to day more hostile to Venus.

In all ranks and conditions of life there are to be found people with extraordinary virility, and others with weak sexual power. It is a matter of course that people who are preoccupied by cares, ambition, or any other unusually absorbing state of mind, have little time to spare for love and sexual affairs. They consequently

---

[1] Max Nordau, Die conventionellen Lügen. Leipzig, 1886, 12 Auflage, p. 123.

have a weaker sexual impulse than others who devote a great part or even the whole of their time to pondering over sexual matters. This is the reason why great workers, whether physical or mental, endure sexual abstinence very easily; but only in case they have been abstinent to begin with. Scholars, for instance, who have only in the course of time taken to study, cannot stand absolute continence even during uninterrupted study. They are seized with vertigo and uneasiness in the midst of the most intense mental efforts, these evils not abating until sexual satisfaction has been obtained. I have had occasion to watch more than one case of this nature, and the patients have always experienced good results by following my advice,—viz., to give up absolute abstinence.

Finally, I must once more refer here to the notion that has descended from Hippocrates to Lallemand and Roubaud. In spite of Hippocrates and his followers, I repeat that *riding* cannot exercise an injurious influence on the sexual power. A single glance given to cavalry will convince to the contrary. Not to mention my own experience with cavalries of different countries, I will point to the fact that where horse-rearing is carried on and horseback-riding is very common the number of births, both legitimate and others, is rather above the average, and the morality of such places is somewhat light. In consideration of all this it is obvious that riding is no hindrance to the development of the sexual power.

SENILE IMPOTENCE.

The energy of the bodily functions diminishes with the advance of age. The sexual function, being one of the last in its development, is also the first usually in the file of the functions that gradually desert the body, growing by degrees weaker. Old men consequently are impotent according to the course of nature, and yet there are very old persons who can still accomplish something remarkable in sexualibus.

We must admit a difference in old people, considering more the condition of the body than the number of years. Appearance alone often deceives, as we may see old persons whose bodies have preserved all their forces but the sexual. The reverse is hardly ever seen, with the exception of those pathological cases where in decrepit old men there is a sexual impulse quite out of proportion. There are perfectly vigorous old men, as there are also quite decrepit young people. Every physician has opportunities of observing men of seventy or more years who possess'excellent power of assimilation of material and also quite energetic sexual functions.

The autopsies of old men made by Duplay, Dieu, and others show that even octogenarians may have *well-developed spermatozoa*, which, however, does not prove that they are virile, but only that they may be so. After all, sexually vigorous old men are *exceptions*, because the functional capacity of the sexual organs generally begins to diminish with the fiftieth year, continuing to decrease until the sixty-fifth year, when it is generally extinct.

As the greatest *individual differences* prevail in this respect, it is quite impossible to set a fixed time for the beginning of physiological senile impotence. Nor is it possible to determine in every case why such or such an individual has grown impotent in early life while another is still perfectly vigorous at an advanced age. It is noticed that in some families a premature impotence and in others a tardy extinguishment of the sexual power is, so to speak, hereditary.. Some individuals who have been healthy and strong all their lives remain sexually vigorous to a good age. Again, other individuals are seen who have always used with prudence their sexual power and possess otherwise the requisite qualities, preserving their manhood to a high age. Too frequent excesses, especially in onanism in youth, and over-careful husbanding of the sexual power are the greatest enemies to the preservation of virility. It is extinguished earliest in individuals in whom it has never appeared with impetuosity, and who, on account of this feeble desire, have acquired renown for virtuousness. It disappears latest in those who may now and then have given rein to their impetuous impulse, but without going in their enjoyment beyond a reasonable measure, —who have, in a word, given off at all times only what they could easily spare.

Only a superficial observer will be surprised at seeing one individual quite impotent when old, after having solicitously spared his sexual power all his life, and another, known as an epicurean, who still possesses a certain degree of sexual vigor in spite of his advanced age. Medical science must not be unfair toward such exceptions, which are frequent enough. It must not thoughtlessly follow the dictum of the past, and deny them every sexual power, together with the right to make use of it. I am aware that these old veterans will not care for the well-intended but strict prohibitions of too scrupulous medical authorities. I know they will, all the same, put into practice their right as much as possible and feasible, and I will frankly oppose my opinion to those scruples and say that old people run no risk in satisfying real sexual wants.

Of course, I am not including here those pathological

cases of an increased or reawakened sexual impulse in old age after it had become extinct. I am speaking here only of preserved sexual vigor in advanced age. I believe that satisfying real sexual wants can be but advantageous to old age, as it contributes to stimulate the energy in the assimilation of material; it buoys up and makes the heart rejoice; it helps to keep up cheerfulness, which is generally reduced in old people, and therefore may properly be considered as a means of favoring longevity.

As to those cases in which death occurred soon after some old gentlemen had entered into married life, or those cases of sudden death before, during, or after coition, nothing is proved by them, since we hear every day of persons dying slowly or suddenly without having thought of marriage or of sexual intercourse for a long time past. Again, we also see men rejuvenated by the side of young wives and living to an advanced age. At any rate, those old men who are still in possession of a good remainder of their sexual power have a better prospect of a long life than those who in a decrepit state are condemned to a virtue which is not always voluntary.

Modern authors are beginning to conform to these ideas. Edward Martin,[1] for instance, knows "of one man who at the age of seventy-eight has begotten a child, and who states that his erections are as vigorous as in youth, and that he performs the sexual act frequently and satisfactorily. This man's powers are possibly kept alive by his marriage with a young and vigorous woman."

---

[1] Impotence and Sterility. Hare, System of Practical Therapeutics, vol. iii. p. 661.

CHAPTER VI.

DIAGNOSIS.

Although the causes are manifold, yet all the forms of impotence have this in common, that the diagnosis is in the first instance based upon the *subjective* sensations and observations of the patient himself. It is well known how unreliable the statements of patients are. *Objectively* there is very little to discover in most of these forms; while in some the external sexual organs show various degrees of flaccidity, shrinkage, and paleness. An endoscopic inspection of the urethra reveals in most cases different degrees of paleness of the mucous membrane, and in nearly all cases that are associated with involuntary loss of sperm various degrees of inflammation of the colliculus seminalis are discovered. Very often the result of such inspection is negative.

Although these few points of diagnosis are of small importance and are not always present, yet the physician must never fail to examine minutely the sexual organs with the endoscope and without it, because the positive or negative result will yield some points in eliciting the causes and helping to determine the appropriate treatment. The objective results, whether positive or negative, together with the statements of the patient, must then form a whole, from which the physician will deduce his opinion, or, relying on which, he will engage in further investigations. The result of an endoscopic inspection is in many cases very important. Unfortunately, many persons are very sensitive to the use of an endoscope, and this is easy to understand if we consider that they are in many cases neurasthenic and timorous. Proper encouragement and confidence in the result will, however, persuade the impotent to submit to almost anything.

When any one consults me upon impotence, I usually begin with close questioning; after this I proceed to the inspection of the exterior of the whole body, and particularly of the sexual organs. This is followed by an endoscopic examination, and, in conclusion, I question the patient about the various points that may have been observed during the inspection.

A physician on the point of engaging in some treatment or other from which he expects any satisfactory result must know regarding his patient the age, hereditary conditions, occupation, constitution, manner of living, previous diseases and sexual life, present state of health, and every detail of the existing degree of impotence. He must, besides, subject to a careful examination the whole body of his patient, and especially the genitals. The penis as well as the testicles and spermatic cords must be examined in regard to size, the amount of blood they contain, and their sensitiveness. The introduction of the endoscope will indicate at once the width and sensibility of the urethra, the color and other conditions of the urethra, and especially of the colliculus seminalis.

The diagnosis is easy in most forms of impotence, but is subject to frequent mistakes on account of the *unreliability of the statements* of the patient. Modesty, ignorance, false notions, excessive timidity, and an inclination to falsehood are the commonest causes that induce a patient to make many a wrong statement wittingly or unwittingly.

There is no difficulty in the diagnosis of *congenital or acquired defects* in the formation of the external genitals. In cases where the congenital defects concern the internal genitals, without any outward signs, some light may be thrown upon the question by endoscopy and by a microscopic examination of the sperm.

In impotence *following some disease* not located in the region of the genitals the diagnosis of the primary or causative disease is sufficient to elicit the cause of the impotence, and then, generally, there are no corresponding pathological changes about the genitals.

In some, though very few, cases of *inherited predisposition* to impotence the diagnosis is very difficult, because the appearance of the patient and also the condition of the visible sexual organs seem to contradict the statements of the patient. Again, in other cases, while the external genitalia may not appear to correspond to the condition of the rest of the body, no sufficient explanation is presented for the complete impotence that exists; hence the physician must depend largely upon the statements of the patient for a diagnosis. The local temperature and electro-sensibility are in such cases of particular importance.

In the *neurasthenic forms* of impotence resulting from abuse of the sexual power the conditions are of very varying nature. In some cases there is absolutely nothing to be established objectively, except a diminution in the sensibility and electrical irritability. In other cases, again, we find abnormal pallor and laxness in the external sexual organs, with or without local inflammation in the ductus ejaculatorii and neighboring parts, always associated with paleness of the urethra. Grünfeld[1] states that in onanists he found hyperemia of the colliculus seminalis nearly constant. The usual signs are dark-red, even scarlet color, and hypertrophy associated with slight vulnerability of the colliculus seminalis. In spermatorrhea Grünfeld found catarrhal swelling of the colliculus seminalis. In high-graded spermatorrhea accompanied by impotence a yellowish-red coloring takes the place of the reddening of the mucous membrane. In individuals suffering from pollutions Grünfeld found in some cases a kind of hypertrophy of the colliculus seminalis. Such results or conditions may be recorded by every endoscopist. The objective findings will vary according to the nature of the abuse that has taken place. In cases induced by excessive onanism we find, without exception, laxness and pallor of the penis and testicles, a smooth scrotum, and low-hanging testicles in consequence of the relaxation of the muscular fibers in the tunica dartos. The orifice of the urethra is reddened, the rest of the urethra pale as far as the colliculus seminalis, and this latter is in different degrees of inflammation.

The vaguest signs are presented by those cases of impotence which have arisen from excesses in venery. In forms of impotence consequent on abstinence the result of examination is only apparently negative, since the testicles always show smaller dimensions than is natural, although the penis may not show any change in its form.

The objective result in the external form of the sexual

---

[1] Die Endoscopie der Harnröhre und Blase, Deutsche Chirurgie. Stuttgart, 1881, Lieferung 51, p. 173.

organs is always of relative importance, because it has first to be compared with the former state before a conclusion can be arrived at about the abuse that has taken place.

Generally negative results are obtained in the objective examinations of the sexual organs in those forms of sexual neurasthenic impotence which have not been caused by a mismanagement of the sexual power, but are founded entirely on a neuropathic predisposition; for instance, relative impotence not induced by weakness never presents any pathological changes. In the other neurasthenic forms,—as, for example, so-called psychical impotence,—we discover here and there signs of atony of the sexual apparatus, and, besides this, in so-called irritable weakness we discover sometimes different grades of inflammation of the colliculus seminalis.

CHAPTER VII.

PROGNOSIS.

There is nothing to be said about prognosis in general, since every individual case carefully considered has first to establish fundamental points on which may be grounded a prognosis that even then is not always reliable.

The prognosis is absolutely unfavorable in cases of absence of the penis, of both testicles, of excessive smallness of the sexual organs, of excessive hypospadia or epispadia. It is more or less favorable in the other kinds of organic impotence.

In the forms of impotence dependent on other bodily defects the prognosis is based entirely on the physician's ability to remove the primary disease.

The prognosis in the forms of congenital impotence is always doubtful, because we seldom succeed in the treatment of inherited defects or abnormalities of the sexual instinct, or in the effort to lead it into a more natural channel.

In the forms of neurasthenic impotence following bad management of the sexual power the prognosis is very varying, depending on the symptoms in each individual case. In cases accompanied by persistent pollutions the prognosis is always doubtful, because we can never know whether the pollutions or spermatorrhea can be mastered. In impotentia ex abstinentia also the prognosis is doubtful in case atrophy of the testicles has already set in. The prognosis is generally favorable in the forms of so-called psychical impotence.

My personal experience proves to me that Eulenburg[1] is not absolutely correct when he makes the statement that the prognosis is better in cases of merely functional injury, in hyperesthesia of the prostata and pars prostatica, than when there are serious and palpable structural changes, such as cystitis, prostatitis, strictures, etc. I would declare this assertion correct only in reference to grave "structural changes," in which case I should, however, not include cystitis or stricture, nor even simple prostatitis. For my part, I much prefer to find those structural changes to be the cause of the impotence or sexual neurasthenia than to discover that the neurasthenia is an independent disease. In purely neurasthenic disorders the successful treatment is not so easily accomplished as in cases of curable structural changes.

The prognosis in general depends on the result of the examination. We can augur it to be favorable in case of youth or, at least, early manhood, strong constitution, otherwise sound general health, preserved sensibility, electrical irritability, normal temperature, and vascularity in the sexual organs. The absence of these conditions makes the prognosis correspondingly doubtful, or even positively unfavorable.

The physician should, however, always bear in mind that there are few persons impotent through any cause except old age who cannot profit by a rational or judicious treatment. Consequently he must engage in the

---

[1] Sexuale Neuropathie. Leipzig, 1895, p. 33.

treatment of impotence with just as much zeal and courage as he would display in the treatment of any other curable disease; and he may feel convinced that by the cure of one impotent individual he will dry many tears and do a great amount of good.

CHAPTER VIII.

PROPHYLAXIS.

" Medicina est conservatio sanitatis et curatio ægritudinis "

It must be granted by all that impotence is one of the modern diseases. A physician who does not occupy himself with impotence has no conception of its great prevalence, nor does he understand that young people may be impotent without any one suspecting it. Very few have the courage to consult a doctor about these *maladies honteuses*, as they are called by all civilized nations.

Only an insignificantly small proportion of those who have become prematurely impotent owe this severe infirmity to an inherited congenital or innocently acquired deformity or disease. The majority have become prematurely impotent because they have been left to themselves and to chance. However well an individual may be led and directed in general, in the most important concern of life, the sexual life, he commonly receives no guidance. If one is possessed of spirit or energy and common intelligence, he may now and then, in his sexual experiences, stumble into different snares; but if no serious injury is sustained, he may eventually attain senile impotence, the natural goal of a healthy man. However, an unlucky accident or the following of a misdirected path may lead to a disease that brings a speedy end to virility.

So, before we can speak of a general prophylaxis of

impotence, we should first feel convinced that it is highly unwise to allow any young man to enter upon the path of physical love, which is strewn with thorny roses, without furnishing him first with some good instruction. It is true, most men acquire a certain experience in time; but, alas! many pay for it very dearly, and often with the loss of their power.

The prophylaxis of impotence is closely connected with the prophylaxis of onanism, because the great majority of all those who have become prematurely impotent commenced with onanism at an early age. First of all is required a strict but loving and rational surveillance of the children. Next comes instruction as soon as the first signs of puberty appear. This instruction, however, should be given without heating the imagination and without the help of those books of horror, the so-called popular scientific works which generally contain some piquant stories.

When a child has given evidence that he practises onanism, every possible effort should be made in order to induce him to desist from the evil habit. The details of the various ways of discovering onanists, the methods for curing the evil with its consequences, etc., cannot be treated here at length. All this may be found in special works, and we wish particularly to mention that of Fournier.

In the treatment of onanism the individual has to be carefully studied: not every child, nor youth, nor even man, has sufficient will-power to combat successfully this evil so difficult to conquer. In many cases the object will be attained by incessant watching, or ultimately by the application of a suitable preventive apparatus, which the child must wear day and night.

Mature individuals should be advised to satisfy the sexual instinct in a natural way, and no notice must be taken of the cry of horror uttered by pharisaical medical authorities or by those who, although possessed of great scholarship, are nevertheless destitute of experience.

The notion that whoever has once enjoyed natural copulation will not feel tempted to return to onanism is an error that is somewhat prevalent. Only copulation that is practised regularly, satisfying every strong and real desire, can cure onanism; while copulation enjoyed at long intervals only would rather incite to more frequent onanism, because pleasing recollections are near at hand.

We know that absolute continence is attended by bad consequences; it gradually extinguishes the sexual power, and does so the sooner and the more easily the weaker the original virility. It is really ludicrous for Bourgeois[1] to admit that he prefers nocturnal pollutions to coition, and to ask why one cannot leave the sexual organs inactive and enjoy good health. He points to the peasant, who does not exercise his mind, and the prisoner, who does not fatigue his apparatus of locomotion.[2] No doubt this may be true, but then we shall see the abstinent just as virile as the peasant is intellectual and as the prisoner is enjoying his exceptional health.

We have already stated that it is utterly impossible to fix a general rule as to how often coition is to be accomplished. Personal disposition and force, phenomena

---

[1] Les passions. Paris, 1877, p. 123.
[2] Bourgeois, op. cit., p. 119.

preceding and succeeding coitus, are proper guides to show each reasonable individual where the line of sufficiency is drawn. Every effort beyond this is injurious. No individual should take another as an example for his own conduct, because constitution, hereditary condition, temperament, age, education, manner of living, occupation, state of health, all are active in establishing differences in sexual vigor, either for the time only or permanently, and it would be perfectly useless to combat these influences. Let each man be satisfied with what has been bestowed upon him.

Timorous patients, and healthy persons also, who see in their doctor their best friend, often ask what part of the day or the night is the most appropriate for coition.

My answer is invariably, that moment which is most convenient and when the sexual desire is most urgent. From an esthetic point of view the evening hours are the time for love. Persons with weakened virility are accustomed to take advantage of the erection in the morning. Some medical men have uttered their veto against this habit; but this veto is unreasonable, since, when the erection is not indicative of a real want, the member slacks at the first movements and coitus cannot be accomplished.

The physician is also frequently asked in what position the act ought to be carried out. Here also a brief answer is all that is due: all positions except the upright are equally advisable from a hygienic standpoint. If the ecclesiastic prescriptions on this subject are disregarded, the most convenient position is the one to choose. Coitus *a parte postica* is decidedly the most natural and favorable for generation. It is also the most convenient way for corpulent persons. This is the mode that is said to be followed always in Australia, because the genitalia of the women there are placed a little farther back.[1] The Jews believed that coition in the usual way produced children who were generally not so good, wise, or talented, and did not give ground for so much hope, as those resulting from copulation *a parte postica*. This doctrine was an abomination in the eyes of Mohammed, and therefore he stated, according to the Hediths (traditions), that the following verse of the Koran had descended from heaven: "The woman is your field; come into your field by whatever way you choose." (II. Sure., p. 25, verse 224.)[2]

In intercourse between passionately amorous beings, one of whom is always the leader, there arise sometimes habits of certain caresses that might yield subjects for contention, and which we choose to designate simply as somewhat piquant. The medical adviser would do well to dissuade from caresses that are rather too piquant, because a man may accustom himself to such accessories, and then, when he is refused them, be impotent for the time. *Sapienti pauca*.

The most varying opinions prevail in regard to copulation during the *menstrual period*. If we compare the menstrual period with the rutting season of animals, some question is aroused as to the advisability of having coition with a menstruating woman, for that period would seem to offer the most favorable opportunity for procreation. On the other hand, however, it should be

---

[1] Ploss, Das Weib. Leipzig, 1885, p. 80.

[2] Der Koran übersetzt v. Dr. L. Ullmann; Nicolaus v. Tornauw, Das moslemische Recht. Leipzig, 1855, p. 73.

stated that most nations observe a custom, a religious rule, or a law, that a menstruating woman is not to be touched. It is further to be observed that some men with sensitive mucous membranes may get urethral catarrh. Again, it is to be noted that many women are actually sick during menstruation; that coitus with a menstruating woman shocks the sense of cleanliness; and, finally, that the increase of the population has not been affected among nations like the Jews and Mohammedans, where the woman is declared unclean during her menstruation and coitus is strictly prohibited. Thus copulation may be desisted from during such period.

In recent time much debate has been carried on about the hurtfulness of a certain process during copulation, which the French have termed "frauding." In consideration of the difficulty one meets nowadays in providing for a large number of children, even married people often feel compelled to accomplish coition with certain precautionary measures against impregnation. For this purpose are used condoms, the Paris or safety sponges, pessaries, and other similar arrangements. Generally, however, the penis is withdrawn just before ejaculation takes place. Only he who has no idea of what it means to have eight children and little or no bread will contend against the justification for these precautions. The use of a well-made and sufficiently elastic condom, a Paris sponge, an occlusive pessary, or similar arrangement certainly has no injurious effect on the man, and, I venture to assert, no harmful effect on the woman. The theory of the cooling of the uterus by the ejaculated semen, advanced by Al. Mayer and Devay, and discussed with such complacency by

Bergeret,[1] has no foundation whatever. It is easy to convince oneself that the woman during ejaculation experiences only a pleasant sensation of warmth and moisture, and by the use of the above-mentioned measures of precaution the continued friction of the penis against the clitoris and the whole surface of the vaginal mucous membrane assists the woman in continuing the venereal orgasm to the end.

The circumstances are far more favorable for the man than for the woman when the penis is withdrawn the moment before ejaculation. For him the erethismus ends with the act of ejaculation, and he does not suffer in any way if a smaller quantity of sperm is emitted in consequence of the premature ceasing of the movements of coition. Matters are different concerning the woman: she is sometimes in the midst of a most intense orgasmus venereus when the cessation of the friction occurs suddenly, and this may cause disturbances in the nervous as well as in the sexual system; for, according to the present state of our experience, it must be assumed that the effect of an abnormal act is injurious,— *i.e.*, if the act has not been continued until the satisfaction of the sensation produced by the ejaculation is experienced. The explanation of this is that if the contraction of the muscles does not take place, the genital tube remains surcharged with blood; the hyperemia subsides but slowly, and may be the cause of changes in the tissue, or a genital derangement.[2]

---

[1] Des fraudes das l'accomplissement des fonctions génératrices. Paris, 1884

[2] Krafft-Ebing, Ueber pollutionsartige Vorgänge beim Weibe. Wiener med. Presse, 1888, Nr. 11.

The statements that Bergeret makes in this respect are probably much exaggerated. I am in a position to make the following statements resulting from personal experience. Some women bear perfectly any kind of frauding, even the last mentioned, whilst others very soon become nervous, and even have hystero-epileptic fits or suffer from catarrh of the cervix of the womb. In contrast with earlier experiences, several cases have come under my observation in more recent years, in which I found that with men also frauding caused some slight neurasthenic phenomena and an injurious effect upon the sexual desire. The sudden interruption of coitus is not easily borne by passionate men, the individual differences, however, being great. It is certainly possible that "the bad habit of withdrawal indulged in with the object of preventing conception of the woman without foregoing the pleasures of coitus" may be one of the causes of prostatorrhea; but I cannot agree with Sturgis[1] when he further reasons that because indulgence in coitus interruptus does not produce the same satisfaction which coitus does, "there is a constant hankering for more intercourse. This inordinate desire gives rise to more frequent copulation, until hyperæsthesia is set up in the prostatic urethra, which is thought to be relieved by more coitus, and thus a vicious circle is established; the more the patient copulates the more the irritation, and the greater the irritation the *more* the desire for coition."

My experience in nearly all the cases observed was a

---

[1] Sturgis, Prostatorrhea simplex and urethrorrhea ex libidine, Journal of Cutaneous and Genito-Urinary Diseases. New York, June, 1898, p. 270.

lowering of the sexual desire in men, and consequently it cannot be the "*over-indulgence* in coitus which does most of the mischief."

Bergeret seems to have conceived a pet idea which he works out. If a man or woman given to such habits is seized by any ·disease, he attributes it to frauding, although every one of the diseases he mentions occurs without any discoverable cause. Bergeret goes so far as to adduce theological reasons against frauding. He would deny marriage, and therefore copulation, to the poor. In spite of daily experience to the contrary, he asserts that a mother of eight or ten children looks young in comparison with a woman who has for a few years only been addicted to sexual extravagances. Bergeret, consulted by unmarried women of different ages who, in consequence of frauding, are declining and suffering from profuse menorrhagia, is able to cure them by advising marriage. As if by enchantment they all become pregnant and well. We can but come to the conclusion that Bergeret rides a hobby, and is ready to attribute to the habit of frauding any disease observed in a person addicted to it, without taking into consideration that all these diseases are to be found without any apparent cause.

CHAPTER IX.

TREATMENT.

Impotence, always difficult to cure, is often incurable. The great number of methods and remedies recommended speaks for the small value of most of them; and yet there is hardly any one of them that could be entirely dispensed with, because there are cases in which the one or the other may be of some use.

The treatment of impotence, this many-headed hydra, varies according to form, phenomena, and state. The treatment of one form of impotence varies in regard to stage and accessory phenomena, and the remedies have often to be changed before one can obtain a cure. One and the same remedy has not an equal effect on all men; allowance must be made for, or due attention paid to, idiosyncrasies of the patients, who are most of them neurasthenics.

In the choice of any therapeutic measure we must take into account everything,—the state of the sexual organs and general bodily conditions, the time taken by individual metabolism, age, habits, occupation, and manner of living. Often the system accustoms itself to a remedy and renders quite inoperative one that had formerly done favorable work.

The success or failure of a treatment depends on the choice of the remedies, and in order to be able to choose the proper one in a given case the physician must, first of all, have a great amount of experience in that direction; he must make his examinations with great care and employ much ingenuity. In order to secure a prospect of success, he must, in the first place, gain the confidence of his patient. The sick, as a rule, approach the medical man with little hope and confidence. To this is added false shame, which makes them very reserved. Most physicians care little for such patients, and therefore dismiss them after a superficial examination. This will not help to increase the hopefulness and confidence of the patient. In order to win this the physician should at least show a certain degree of interest and sympathy. He must question his patient very closely and examine him equally carefully. This is necessary for the diagnosis alone.

If the physician inspires the patient with some courage and confidence, he has by that means taken the first step in the treatment, because every impotent person must first of all be treated psychically. As soon as such a patient has once conceived some hope for his curability and confidence in the physician he shows himself an exemplary patient and will subject himself to any treatment. Nothing is too difficult, nothing too disagreeable, nothing too painful. Even individuals who are very much reduced in strength and energy, in consequence of onanism or pollutions, are no exception in this. I emphasize this expressly in contrast to Lallemand's [1] statements, which may have been correct, because in his time the sick had more cause to fear medical men and their methods of treatment.

In case the physician should discover some of the

---

[1] Lallemand, op. cit., tome iii. pp. 129 and 131.

causes of impotence still in existence, he must *remove* them as speedily as possible. Often the impotence disappears with its causes; but at any rate there can be no question of any treatment for impotence before the causes are removed.

The methods for treating impotence are manifold, as already stated. There is a general and a local treatment; the application of medicaments, hydro-therapeutics, electricity, and massage. Each one of these groups comprises many single methods and remedies. In the treatment of a disease that is so difficult to cure we must make use of every means at our disposal, each remedy being applied at the indicated moment and opportunity. It goes without saying that a mere excitation of the genitals or their nerve-centers which would be but transitory in its effect, and would not at the same time have the tendency to strengthen them, can never be the object of a rational and conscientious therapy.

We shall now discuss every curative method, and conclude with the discussion of the therapy for every state of impotence.

Psychical treatment is indispensable in every form of impotence excepting the organic. Psychical treatment forms in some measure the introduction and beginning of every other manner of treatment. It has been stated above that first of all the physician must *conquer the hopelessness and distrust of his patient.* This is frequently very difficult of accomplishment, particularly with patients who have already engaged in the study of several so-called popular scientific works, and with the greatest difficulty with patients who happen to be medical men. I have had many an opportunity to treat neurasthenic physicians for impotence, and especially since the appearance of the first German edition of this work. I met, without exception, with nearly insurmountable difficulties.

The next step is to induce the patient not to think continually of his disease. For this purpose those who are impotent or believe themselves so should endeavor to find various *distractions.* They should be recommended any kind of pastime which involves some bodily exercise; such, for instance, as some suitable occupation that is not fatiguing and is at the same time attractive,—driving, riding, theaters, concerts, balls, gymnastics, fencing, swimming, skating, bicycling, rowing, pleasant journeys that are of moderate duration, etc.

The patient should be most strictly *forbidden* any *useless sexual excitement,* reading of lascivious books, contemplating piquant pictures, and so-called mental onanism. Persons who have experienced a repeated fiasco with women are in the habit, before they proceed to coition, to excite themselves sexually in sundry ways in order to prepare themselves for the act; but they regularly find that the result is the very opposite to what they had in view, and this introductory excitement is often the cause of impotence.

The patient must be told not to allow single failures to affect him too gravely, but to look upon them with more indifference, and treat them as casual mishaps. It is not uncommon that virility returns with the *peace of mind,* while erection will not appear when it is most ardently wished for. This is a reason for the well-known fact that young husbands who fancy they are impotent are often cured by the mere forbidding of coition. The object is to re-establish mental composure, with which often comes the erection also.

A physician must never suppose that it is possible to put an end to impotence by simply denying it, even were it psychical impotence. If the physician denies facts that have been experienced by the patient, he simply loses the latter's confidence irrevocably. Frequently medical men are led to believe they have to deal with a case of hypochondria, when upon investigation they find they have made an error in diagnosis. Hypochondria is a very rare disease, and exists without any reason or cause whatever only in persons of unsound mind, and even then such mental disturbance is itself the cause of the hypochondria.

We have already said that *any existing cause* of impotence must first be *removed* before the treatment of the impotence itself can commence; organic obstacles must, if possible, be removed by surgical operation; diseases causing impotence should be treated appropriately; onanists must be cured of their habit, and proper remedies applied in case there is spermatorrhea, pathological irritation, or a condition of weakness in the genitalia.

We shall devote a more detailed discussion to the **treatment of spermatorrhea**, because spermatorrhea is frequently the sole cause of impotence, and plays an important part in nearly every case. The treatment of spermatorrhea is truly in a lamentable state. The first increase in the frequency of pollutions is scarcely ever treated rationally, because the patient either does not mind it, or avoids consulting a physician about it from false modesty, or, finally, because he finds that his doctor neither understands his ailment nor even listens to his story. Yet one needs no demonstration in order to see that it is of the utmost importance that spermatorrhea should be combated from its very beginning. In our practice we often discover what desperate efforts are made by the patient in his struggle against the constantly increasing frequency of the pollutions; what unreliable remedies are adopted, and how the body is chastised; or how every enjoyment, every comfort, is denied oneself, everything is tried that has been mentioned or praised by physicians, friends, and books as being of good effect in the case in question. Many a remedy or measure seems at first to have some good effect, but loses all efficacy after a time. Meanwhile the precious time passes and the pollutions grow worse. Lallemand[1] tells of some examples that are really characteristic and are taken from life.

The treatment of spermatorrhea is a very delicate and difficult affair. It engages in its service nearly all the therapeutic remedies and expedients which find their application in the treatment of impotence, and of which we shall speak later on. Thus it includes in some special cases the use of medicaments, hydro-therapy, electro-therapy, and local endoscopic treatment.

As a matter of course, we must in the first place search for the etiological factor, so that it may be removed as soon as possible. Onanism, which is the most frequent cause of spermatorrhea, must be considered at once. Besides this, a phimosis, if such exists, must be corrected by surgical operation, even if it is not of a high degree. Phimosis is oftener the cause of persistent pollutions than one would think, because by pressure it irritates the member, even when slightly erected; and, besides, the phimosis, by protection, renders the glans more irritable or sensitive than is to be desired.

---

[1] Pertes séminales, tome i. pp. 294–304.

Regulation of the manner of living will present special difficulties. An individual suffering from pollutions must in the evening abstain from food that is difficult to digest; he must not eat any spicy dishes, and in general special attention and watchfulness must be directed to the food and the increase of the activity of the digestive organs. The patient must take his supper at least three hours before sleep, and he must before he goes to bed empty his bladder, and at all times care for regular defecation. He must sleep on a couch or bed that is moderately hard, but not too hard,—best on a horse-hair mattress. My observations have taught me that persons suffering from persistent pollutions will be benefited by sleeping with the head in a very low position, so that the brain can be better fed by blood.

Again, the patient must not sleep longer than the necessary time, which should be determined for him. In bed he must not be covered too warmly, and he must not sleep on his back. His trousers must not be too tight. He must never sit on upholstered seats; he must not ride on horseback; he must go as little as possible in a conveyance; and he must not excite himself sexually without necessity. Of special importance is the regulating of sexual intercourse, because the pollutions cannot possibly be cured during absolute abstinence.

In the treatment proper the physician has to pay strict attention to the individual case. Hydro-therapeutic measures combined with a well-regulated manner of living, and eventually a trip to or a sojourn at a watering-place, will constitute a curative method that most frequently leads to a satisfactory result. Balneotherapy[1]

---

[1] Kisch, Grundriss der klinischen Balneotherapie. Wien und Leipzig, 1883, p. 292.

gives the best result in those cases of pollutions and spermatorrhea in which onanism and weakness of the nervous system are the causal factors. The use of pure chalybeate waters and ferruginous waters charged with carbonic acid is indicated in such cases.

If the pollutions have originated from a state of hyperemia in the pelvic organs caused by some abdominal stasis or habitual constipation, good effect will be obtained by taking waters containing sodium sulphate, sodium chlorid, then the different bitter waters and some sulphur waters. We might add here that in a state of irritation in the sexual organs—*i.e.*, in cases where chronic inflammation of the mucous membrane of the bladder or of the urethra is the cause of pollutions—the alkaline waters are to be recommended, and will do excellent service. When the mineral waters are taken internally we should observe certain precautions; for example, only small doses must be ordered, so as not to surcharge the bladder, whereby an irritation would be induced. The waters must not be taken in the evening.

When an increased morbid sensibility of the nerves is the cause of the pollutions we recommend the acratothermæ of an elevated region; and the iron mud-baths are advisable for cases in which simple anemia is the cause of the disease.

Various cold water treatments and sea-bathing are indicated in most conditions of weakness causing spermatorrhea, but these remedies should be selected cautiously, individual characteristics being taken into careful consideration.

Rational gymnastics will always be of great value in

subduing frequent pollutions. Schreiber[1] recommends for this purpose different exercises for chamber gymnastics which are very easy of execution, do not require any apparatus, and are of very good service, particularly so when supported by some other means of treatment.

An electrical treatment is only rarely indicated. In the endoscopic examination some local treatment will often appear to be required or necessary, mostly the use of the sound or bougie and Lallemand's method of cauterizing; but always by the guidance of the endoscope, of course. These measures lead most frequently to the desired result.

Bromids taken internally are sometimes of excellent effect in cases accompanied by erotic excitement, yet they are not infallible. Prescribing them is an easy matter, and the sight of a prescription is always some satisfaction for both doctor and patient; only this satisfaction is not always of long duration with the patient. Camphor also may be used in some cases, but oftener in the shape of a suppository than otherwise. A trial with secale cornutum, with tinctura veratri viridis, monobromated camphor, antipyrin, sodium nitrate, and eventually with solutio Fowleri, may also be advisable.

According to Rosenthal,[2] atropin has a good effect, but only in cases of prostatorrhea. I have a prejudice against atropin, and thus have never ventured to make any use of it, although in certain cases I have felt tempted to make an experiment with it in consequence of Loewenfeld's[3] recommendation.

---

[1] Aerztliche Zimmergymnastik. Leipzig, 1883, p. 95.

[2] Ueber den Einfluss von Nervenkrankheiten auf Zeugung und Sterilität. Wiener Klinik, 1880, Heft 5, p. 160.

[3] Die nervösen Störungen sexuellen Ursprunges. Wiesbaden, 1891, p. 158.

M. Meisels, under the direction of Professor A. Bokai, made some experiments with cornutinum citricum, and he asserts that doses of 0.003 to 0.006 (gram) per day acted very favorably in paralytic spermatorrhea. In most of the cases the sperm effusion diminished on the second or third day, or it ceased altogether. In from one to two weeks a cure generally resulted. This medicament had no disagreeable effects even after nine months' continuous use. In spastic forms, however, cornutinum citricum is considered to have no effect. I have tried this remedy in one case only, with apparently good effect. The price is excessive.

Professor Bozzolo and Mangianti recommend the following prescription for spermatorrhea and anaphrodisia of neurasthenics: R. Cornutin. citr., 0.03; cretæ præpar., 3.0; gummi tragac., 6.0. M. f. pil. No. xx. S. 2–4 pills daily.

Of course, diseases which induce pollutions must be treated and removed in any case whether pollutions appear or not. Such are ascarides, itching and smarting cutaneous eruptions about the genitalia and vicinity, also hemorrhoids and fissures, strictures and phimosis.

The various mechanical devices invented for preventing pollutions cannot be of much use, since, though, if well constructed, they may prevent now and then a nightly effusion, they do not thereby contribute much toward the complete removal or cure of the disease, except in certain cases of neurasthenic conditions and states of habit. The arrangements that prevent the patient from lying on his back deserve more consideration.

In special cases I have followed Lallemand's [1] exam-

---

[1] Lallemand, op. cit., tome ii. pp. 46–56.

ple and ordered continuous application of cold, whereby some rather satisfactory results were obtained. I used in these cases Chapman's tubes or pipes, which facilitate the application, which is not very convenient under any circumstances.

Kisch[1] states that partial bathing is recommended against pollutions in youthful individuals. For the bathing of the occiput, the patient is in a horizontal position and has the back of his head in a specially shaped basin filled with cold water. Stimulating applications for the upper arms, by means of some towel-like material soaked in cold water, and with a dry cover, are likewise of good effect now and then. Finally, I wish to mention for occasional application Winternitz's psychrophor, cold clysters, and Atzperg's cooling probe for the rectum.

Sometimes the cure may be assisted by the patient's will and firm determination to awaken at the proper time.[2] For the cases where the pollutions take place in the mornings, we have L. Casper's[3] advice to awaken the patient regularly by some arrangement an hour before the phenomenon usually occurs, so that he may urinate. This is certainly good advice, because in this way you may break the force of habit. The patient must not then be allowed to fall asleep again.

I shall now return to the discussion of the different modes of treating impotence itself. In sexual weakness or anaphrodisia special weight is to be laid on a **hygienic manner of living**. Food, physical exercise, and rest,

---

[1] Op. cit., p. 293.

[2] Campbell Black, On the Functional Diseases of the Urinary and Reproductive Organs. London, 1875, p. 172.

[3] Dr. Leopold Casper, Impotentia et Sterilitas virilis. München, 1890, p. 102.

also dwelling and clothing, must be strictly directed according to the rules of hygiene.

The food must be nourishing without being stimulating. Roubaud has with great industry compiled a long list of so-called *aphrodisiac articles of food*, etc., which may be of some benefit in special cases of frigidity and psychical impotence, or for the days preceding an intended copulation. These articles are in some measure of mild effect, sometimes of no effect, but certainly are harmless stimulants. Here is the list: Salt, zedoary, saffron, mustard, cinnamon, sage, rape-plants, as carrots, turnips, etc., marjoram, nutmeg, cardamomum, mustard-plant, arrowroot, laurel, leek, ginger, garlic, onions, cloves, peppers, skirret or sugar-root, cryngo, angelica, parsnips, celery, fennel, vanilla, pork, game, oysters, fish, etc.

In general, the nourishment must be suited to the state or condition of the body, every superfluous production of fat being injurious to virility. On the other hand, it is to be noticed also that individuals who are possessed of considerable sexual power enjoy a good appetite and digestive power, although they are not gourmands, or do not become such until their riper years. Persons with low sexual capacity are either gluttons or possessed of small digestive power.

The diet is of special importance in the treatment of all chronic diseases. It certainly influences in a certain degree the virile power, and every physician will therefore do well to inquire about the nourishing material of every impotent patient, and correct any existing error. Our purpose does not admit of the use of the diet for producing corpulency, as indicated by Mitchell, Playfair, and others. I am ready, however, to admit that considerable good may be accomplished when such treatment is modified in the manner that Fürbringer[1] proposes,—*i.e.*, that the patients do not have to stay in bed, are allowed open air exercise and light mental work; but then we do not follow Mitchell and Playfair. Fürbringer is approaching my point of view, although he does not seem to notice it.[2]

Experience, which was gathered principally in North America, leads me to establish and to follow these main rules. The impotent must abstain from spirituous beverages. I make an exception only in the case of persons of feebly developed sexual desire; these may take two glasses of German beer or one glass of good, strong California wine shortly before intercourse. I have already mentioned the favorable effect of beer in cases of precipitate ejaculation. Beer or wine, however, must never be taken in such quantities that the stimulating effect may be followed by a paralyzing influence, be it ever so slight.

The manner of living must be strictly ordained in accordance with hygienic laws, and a proper proportion observed between physical or mental occupation and rest, which, however, is of equal necessity for the healthy and for the sexually weak. The patient ought to divert himself with mental exercise and amusements of every kind. Gymnastics, walks, and so forth should be resorted to in order to strengthen the body. Fatigue of every kind must be avoided, and every effort must be followed by an appropriate interval of rest.

---

[1] Die Störungen der Geschlechtsfunctionen des Mannes. Wien, 1895, p. 66.
[2] Ibidem, p. 136.

TREATMENT. 239

The principal rest is taken during *sleep*, and this must be apportioned to every individual according to his requirement. The patient in this respect is often in an unfavorable situation: if he sleeps sufficiently long to give his body the necessary repose, pollutions take place during the latter part of this time; and if he denies himself part of this repose, then faintness and exhaustion appear and exert an unfavorable influence on the progress of the cure. We cannot fix upon the number of hours necessary for sleep in every case; but, on an average, eight hours should suffice. Of course, we must pay due attention to the condition of dwelling and clothing.

Although **medicaments** are not the means that in the treatment of impotence lead with great frequency and safety to a fortunate issue, yet they are what every sufferer desires and often asks for, after having given only superficial statements about his complaint. The question, "Will you prescribe something for me?" is never missed. The prescription may sometimes, in connection with other remedies, have a good effect. Thus I shall now proceed to the discussion of the most common medicaments.

Philters, or love-potions, have been known as far back as the times of Moses. The mandrake that Rachel is reported to have eaten to become prolific is now supposed to be atropa mandragora, and belongs to the genus belladonna. Philters are brewed and drunk in our days, with and without effect.

Cantharides and its preparations were in former times the commonest remedies used for impotence. If taken internally, the most effective substance of the cantharides, the cantharidin, is excreted by way of the urinary passages. On these passages cantharidin has a very irritating influence, and in proportion to the size of the dose it may lead to serious hyperemia of the mucous membranes of the urinary passages, to albuminuria, hematuria, and cystitis, and, in the worst cases, to croupy deposits on the mucous membrane of the bladder. As secondary symptoms we may have dysuria, stranguria, and painful erections. These erections are certainly to be called pathological, and yet they are to do service in coition. Doses of cantharides so small that they do not cause any perceptible inflammation of the urinary passages do not cause erection, and doses so large as to cause energetic erections are creative of such dangers that only the despair of a patient or the ignorance of a physician can give rise to a thought of applying them. The patient, sufficiently ignorant and sometimes in such a state of mind as to be willing to sacrifice his life for one night of pleasure, is excusable, but there is no excuse for a medical man who would use cantharides. It is to be hoped, however, that such cases do not occur nowadays. This remedy must never be brought into requisition in the treatment of impotence, nor the *oil-beetles* (May-worms), meloes majales, which are related to the cantharides by the acid of cantharidin which they contain, or the *oil or tincture of ants*, used in South America.[1]

Phosphorus was known to the ancients as a remedy for sexual weakness. It has a stimulating influence on the nervous system. In regard to this remedy we have the observations by Alphonse Leroy and Bouttotz,[2] and those of Delpech, which show phosphorus to be a pow-

---

[1] Rosenthal, op. cit. Wiener Klinik, 1880, Heft 5, p. 163.
[2] Roubaud, Traité de l'Impuissance. Paris, 1876, p. 133.

erful stimulant. I have always had occasion to use phosphorus for impotence, and I feel justified in saying that in most cases it has given satisfaction. I noticed particularly the favorable effect it had on the mood of the patients, and I feel convinced that it is of decidedly good effect in cases where the patient has become indifferent or melancholy. There is no bad effect noticeable in the cautious administration of phosphorus, even from continued use (in pills or capsules of 0.001 gram three or four times daily, or of phosphoric acid twenty to thirty drops in a glass of sweetened water several times a day).

Nux vomica and its preparations are of very great value in all forms of impotence, although the effect is not very vigorous, and, I am sorry to say, is of but short duration.

The extract and the tinctura nucis vomicæ, also strychnin, are justly considered as tonics and remedies that excite the appetite and preserve their character as good nervines in many neurasthenic diseases. Hence it will be worth while to give them a trial; it will be accompanied with some advantage. I saw the best results with patients who were otherwise healthy, but felt a diminution in their sexual power without any assignable cause. It is true, the effect of nux vomica is not lasting, but after it has been used the sexual power will never sink below the level that existed before the use. The doses recommended are from five to twenty-five drops of the tincture, or 0.01 (gram) of the extract, three times a day.

I have never made use of *brucin*, because it is so very unreliable and of such different effect on different individuals.

Secale cornutum and its preparations are also recommended for impotence, or rather as aphrodisiacs; but, as their effect is of but short duration, and, moreover, is quite unreliable, we can dispense with them in all cases unaccompanied by spermatorrhea.

Ergotin is recommended by Maximilian v. Zeissl[1] to be used in combination with quinin. I think that the quinin as a roborant in this combination is the more powerful ingredient in the prescription.

Quinin, either alone or in combination with easily assimilable preparations of iron, will effect in anemic and weakly impotent persons all it is capable of effecting in anemia and weakness. If, therefore, a physician sees some reason for assuming that a patient requires a roborant, then he will with some ground appeal to quinin or iron; but he need not expect therefrom a specific effect on the sexual functions.

Equally uncertain are the volatile stimulants, as musk and castoreum. They may induce libido but no erections, and are consequently dispensable.

In the Orient especially some *narcotics* enjoy the renown of being aphrodisiacs. Indian-hemp, opium, and morphin given in certain doses produce undoubtedly sexual excitement followed by powerful erections. It is a known fact that hashish-eaters and opium-smokers experience heightened sexual impulse in the beginning of these fatal habits. The "just, subtle, and mighty opium" is capable of rousing the sexual desire to a very high degree,[2] and this is, no doubt, due to increased

---

[1] Ueber die Impotenz des Mannes und ihre Behandlung. Wien. med. Blätt., 1885, Nr. 16.

[2] Paul Bonnestain, L'Opium. Paris, 1887, p. 493.

reflex irritability of the spinal cord. These means are, nevertheless, quite unsuited to our purpose, on account of their transitory effect as well as the subsequent relaxation, and also on account of the danger that their use may lead to a habit, the fatal consequences of which are well known. At best the trial might be made to raise the confidence of a patient suffering from neurasthenic impotence; but this would be done at the risk of a perfect failure, since the effect of the opiates is so much dependent upon the individual.

Valerian has unjustly the name of an aphrodisiac, because it only lowers the reflex irritability of the spinal cord, and for that reason it is recommended as a sedative by Arndt.[1]

The *mildly working stimulants*, as vanilla, cinnamon, galanga, and several spices, are of very transitory and unreliable action, and operate only in persons who are easily excited sexually.

Cocain taken internally invariably produced sexual excitement in a man fifty-six years old. I had previously noticed a diuretic effect of cocain, but I am unable to decide whether there is any causal connection. Cheerfulness is always the effect of an internal use of cocain. This is diametrically opposed to the observations of Dr. H. Wells, of the United States Navy, who asserts that he has noticed in cocain an anaphrodisiac effect. Further experiments and investigations would certainly be interesting.

Finally, we mention the *scincus marinus*. This little animal, anything but pleasing, has for a long time been praised as a popular remedy, and in some countries is

---

[1] Neurasthenie. Wien und Leipzig, 1885, p. 246

even now named as a domestic remedy, and yet it contains no substance whatever that can act as an aphrodisiac. At best its fat might possibly induce salacity.

Damiana (Turnera aphrodisiaca), its liquid extracts, and other quite elegant American preparations are not what they are represented to be by the extraordinary advertisements.

For the sake of completeness, and also as a curiosity, I must state that homeopathy also has taken an interest in sexual weakness, and endeavors to treat it by administering the salts of copper, gold, iron, lead, etc.[1]

The various hydro-therapeutic processes have always enjoyed a special fame, and are highly recommended by men in and out of the medical profession for the states of sexual weakness. The reputation of **hydrotherapy** in general increases every day, and with it also that of its special application in the treatment of impotence. Every one who feels a beginning of impotence resorts, with or without medical advice, to cold water ablutions and sitz-baths. These remedies are of feeble action and produce but little effect.

In order to understand the action of water on parts of the human body which are in a pathological condition, we have to call to mind the principles of hydrotherapy. The stimulating effect of water upon the body is always twofold,—thermal and mechanical, the one or the other prevailing according to the manner of application.

The impress of the thermal stimulus[2] upon the peripheral terminations of sensitive cutaneous nerves is trans-

---

[1] Dr. Christof Hartung von Hartungen, Ueber virile Schwäche und deren Heilbarkeit auf inductivem Wege. Wien, 1884.

[2] Winternitz, Hydrotherapie. Wien, 1877, Band i. p. 49.

mitted to the central organs, appreciated by these as sensations of warmth or cold, and transmitted by them by reflex action to the motor system. It is probable that thermal effects have also a local action through the influence of peripheral ganglia or the excitable tissue itself, and without the mediation of the central nervous system. Again,[1] the application of a lower temperature over large vascular trunks causes the latter to contract. This narrowing of the main vessels induces diminution in the afflux of blood toward the peripheral ramifications of the trunk-vessels that are contracted, whereby is also induced a lowering of the temperature in the parts of the body supplied by these blood-vessels. Again, experiments[2] have proven that by the local application of water of different temperatures we can alter at will the local warmth of a part of the body even to the deeper tissues. Finally, stimulation by cold increases considerably the tension and tonicity of the smooth and striated muscle acted upon as directly as possible.[3]

The action of the various procedures in hydro-therapeutics is therefore directed first on the nerves, by means of these on the vessels, and ultimately on smooth muscles. The water must therefore be applied in accordance with the various requirements of the cases under treatment.

There is no case of impotence where one or the other hydro-therapeutic process would not considerably assist any course of treatment; and in many a case no other remedy is required to effect a cure. Again, it

---

[1] Winternitz, op. cit., Band i. p. 75.
[2] Ibidem, p. 36.
[3] Ibidem, p. 119.

must be remarked that an untimely application of a hydropathic stimulus may also do harm.

Out of the immense treasury of hydro-therapeutic procedures we can appropriate for our purposes local and general ablutions, rubbing down, flapping,[1] sponge-baths, rain- or douche- or shower-baths, sitz-baths, half-baths, full-baths, vapor-baths, river-baths, sea-baths, and many mineral baths; also the application of cooling sounds and injections of cold water into the urethra and rectum.

The action of simple *ablutions* is too feeble to be of much good in any form of impotence. They should, however, be observed as a hygienic rule for cleanliness, both by the virile and the impotent. Ablutions of the spine and loins act, nevertheless, as a gentle stimulus. Washing of the spine and genitalia with spirituous fluids is common as a domestic remedy. Roubaud recommends washing with tinctura nucis vomicæ. As a matter of fact, I have seen really good results from the external use of tinctura nucis vomicæ in several cases of purely neurasthenic impotence.

Rubbing down and *flapping* are of excellent service, as they gently stimulate the nerves and assist in the assimilation of material; they thus are indicated in several forms of impotence.

---

[1] I would designate by the name of "flapping" a very important and efficacious hydropathic procedure, which is executed in the following way: a coarse sheet, wet and cold, is wrapped around the body of the patient, the attendant slapping more or less gently, but always rapidly, the whole body up and down repeatedly until the skin is quite reddened and warm. A cold cloth is placed on the patient's head to avoid possible congestion.

Flapping is usually followed by a cold half-bath.

TREATMENT. 247

Sponge-baths may be substituted for shower-baths. They do not operate quite as powerfully, I admit, but they are more easily procured, as a round vessel not overlarge,—a sitz-bath tub, for instance,—or a portable rubber tub and a good sponge are all that are requisite.

Rain- and douche-baths are in many cases absolutely indispensable. Applied generally, the shower-bath assists powerfully in the assimilation of material. Applied locally, on the genitalia and spine, they operate as a gentle stimulant, acting directly on the nerves and spinal cord. In certain conditions the douche filiforme,[1] or thread-like shower-bath, directed upon the glans is of good effect.

In commonest use, however, are *sitz-baths*. Winternitz is of the opinion that the sitz-bath operates by means of a reflex stimulation of the nervus splanchnicus.[2] He has made experiments[3] with these baths and obtained the following results: A short sitz-bath, of ten minutes and 10° C., causes a lowering of the local temperature, which, however, is followed by increased warmth within half an hour, the reaction which ensues during the second half-hour being followed by a moderate decrease in the temperature for several hours. A sitz-bath of thirty minutes and of the same temperature causes a diminution of the temperature during a longer time and to a lower degree. The reaction sets in later, seems less intense, and is followed by a marked compensatory lowering of the temperature. Very long continued and very cold baths might postpone the reaction still more,

---

[1] Winternitz, op. cit., Band i. p. 35.
[2] Ibidem, p. 224.
[3] Ibidem, Band ii. p. 139, etc.

and possibly prevent it altogether. Either short or prolonged sitz-baths of a temperature approaching that of the blood warm the rectum directly. The most important therapeutic results are obtained by hip-baths of 20° C. They usually cause no subsequent warming of the rectum, but constantly show a lowering of the temperature in the rectum.

Hence short, cold sitz-baths must be considered as an exciting, stimulating form of bathing, whilst cold sitz-baths of longer duration induce depression and retard the process of local nutrition and heighten the vascular tonicity in the pelvic organs. Warm and hot sitz-baths have a relaxing effect. Temperate sitz-baths, 18° to 25° C., are antiphlogistic. The cold sitz-bath is of main utility, and should be of a shorter or longer time according as a stimulating or sedative effect is intended.

Of similar but more powerful effect are the so-called *half-baths*, which, along with the ordinary cold baths, must be considered as exciting, if they are not of too long duration, which is not likely to be the case. Very excellent results for the purpose of curing several forms of impotence in which stimulation is indicated are obtained by half-baths of 12° to 18° C., combined with friction and showers during bathing.

Even *vapor-baths* may be indicated in many cases of impotence; these have also a stimulating action, and, moreover, prevent the formation of adipose tissue.

The most stimulating form of bathing is *river-bathing* and *sea-bathing*, which often perform real miracles with the impotent. Milder forms of impotence are very frequently cured by river-bathing alone, and still better by sea-bathing. In these baths there is, besides the thermal stimulus, an exceedingly strong mechanical excitation by the action of the flowing water and of the dashing of the waves. For river-bathing are to be preferred those rivers or parts of rivers which present a moderate depth combined with a strong and rapid current, and, likewise, for sea bathing, places where the billowing is strong. Very great care must be taken in prescribing these forms of bathing, because patients that are run down and many neurasthenics cannot endure them very well. Bathing in stagnant waters, as in *lakes*, has also a good effect sometimes. It may be a thermal stimulus, and the movements made in the bath contribute to the acceleration of the elaboration of matter. Protracted bathing in warm water is scarcely ever desirable.

The *balneological treatment* is merely in its inceptive stage; but we know that soda and sulphurous waters do very good service in nearly every form of impotence. The sulphur waters were recommended by French authors as early as Lallemand's time. I have occasionally noticed a favorable influence on the sexual life of patients who used sulphurous thermæ for other causes. I have frequently ordered with best results bathing in natural salt-water, and also in artificial mineral salt or rock-salt water.

Bathing in other mineral waters is advisable for those forms of impotence in which some specific mineral water treatment will remove causes of impotence, as, for instance, prostration, anemia, torpid digestion, etc.

The Winternitz cooling-sound, *psychrophor*, is an unfenestrated catheter à double courant. Its application is adaptable in the treatment of all the forms of impotence connected with hyperesthesia of the urethra, especially of the colliculus seminalis. The instrument is introduced as far as the neck of the bladder, and sometimes into the bladder. A continuous stream of cold water flows through it, so that to the mechanical stimulus of the sound is added the effect of the cold. The psychrophor is borne even by patients in whom the sound cannot be introduced at all, or not often enough, on account of the violent pain produced.

Simple *injections of cold water* into the urethra are very efficacious on account of the mechanical stimulation and the effect of the cold combined. The effect of such injections is merely exciting, whilst the psychrophor may, according to the duration of the application, have a depressing, even an antiphlogistic, action. The injection of cold water into the urethra is made use of, as I have noticed, by sailors as a means of temporary excitation after long continence, and I wonder that we never read of this practice in medical literature.

In order to act upon the prostatic part of the urethra we may in some cases make use of *Azperger's rectal cooling sound*[1] or of *Winternitz's rectal cooling pouch*.[2] The action is similar to that of the psychrophor, whilst cold-water injections into the rectum act as do those made into the urethra.

Application of *dry warmth or cold* has a similar action to that of hydro-therapeutic stimulation. Very considerable stimulation can be obtained, particularly if high and low temperatures are applied alternately. Roubaud[3] recommends a syringe for the application of hot air.

Many newspapers have advertised the *carbon-douche*. In the application of this remedy the genitalia are, by

---

[1] Winternitz, op. cit., Band ii. p. 129.
[2] Ibidem, p. 131.
[3] Traité de l'impuissance. Paris, 1876, p. 146.

means of a peculiarly constructed apparatus, exposed to the direct action of carbonic acid gas. This action, though mildly stimulating, I consider has more of a psychical effect or influence. The external application of carbonic acid gas has already been recommended by Bernatzik,[1] and quite recently by B. Schuster in Nauheim.[2] The latter found it highly effective in cases of neurasthenic lessening of libido and erectility and in premature senile impotence. I saw very little effect in the few cases in which I have made use of it. Carbonic acid may have a better effect in female diseases.

The different kinds of **electrical currents** are made use of in the treatment of impotence quite as often as the various procedures of hydro-therapeutics. Every kind of current has its advocate among the different authors, and every one extols the method he uses. This fact alone would be enough to prove that science has not yet reached its zenith in reference to the application of these electrical currents.

Erb[3] states in plain words that nothing is known about the electro-physiological action upon the testicles and vasa deferentia of the living man, and that the knowledge of the effect on the spinal cord is also very scanty. We must rely entirely on empiricism, or practical experience, and this teaches us that electricity, in whatever way it may be applied, is of excellent service in special forms of impotence, but that there are many

---

[1] Aphrodisiaca, Eulenburg's Real-Encyclopädie. Wien und Leipzig, 1885, Band i. p. 614.
[2] XVII. Versammlung der balneol. Gesellsch. in Berlin, 1896.
[3] Elektrotherapie. Ziemssen's Handbuch der allgemeinen Therapie, Band iii. p. 128.

cases where it is of no use, or may even do harm. A careful distinction of the sundry cases and a thorough investigation as to which are proper for electrical treatment is more necessary than in any other mode of treatment. Only a thoroughly correct application can sufficiently reward our efforts by good results, and it is of the utmost importance to know what kind of current is to be chosen, of what strength it should be, and in what way applied. This necessitates a careful study of each individual case, with all its accompanying details or circumstances.

The *galvanic current* will be indicated most frequently. We commence by localizing the electricity, applying the zinc pole over the cord in the lumbar region, and the copper pole to the upper and under surfaces of the penis, to the testicles, perineum, and the spermatic cord downward from the inguinal ring. In other cases, when the spinal cord is to be operated upon, the copper pole is applied to the nucha, and the zinc pole to the region of the lumbar vertebræ. A more powerful action is obtained if the copper pole is applied to the lumbar region and the zinc pole to the perineum or to the pars prostatica by means of the bladder rheophore. A still stronger effect is produced by the introduction of the copper pole into the rectum by means of the rectal rheophore, and the zinc pole as far as the pars prostatica. Only weak currents, however, can be applied, and but once a week, as a more frequent application might induce inflammation of the mucous membrane.

In the use of the bladder rheophore Lewandowski[1]

---

[1] Elektrodiagnostik und Electrotherapie. Wien und Leipzig, 1887, p. 410.

recommends the gradually increasing faradic current or very short interruptions in the application of the current. Besides, both poles may be applied externally along the course of the spermatic cords, when interruptions and reversing of the current are particularly effective. It must always be borne in mind, however, that an energetic electrical treatment can be of service only to individuals of very low electrical irritability.

The time of application will have to be longer or shorter as the different cases may require. If the patient is not affected with anesthesia of a high degree, or if the electrical irritability begins to return during the treatment, erections may occur even during the application, and this would very much raise the courage and confidence of the patient.

The manner of application of the *faradic current* is the same as that of the galvanic; but the induced current is not so frequently used for electrizing the spinal cord itself. By means of the metallic brush the glans and the testicles are directly excited rather strongly. Such an application produces a reddening of the skin, and serves therefore as a stimulant to the circulation of the blood in the parts in question to a higher degree than any other method of applying electricity.

If one pole is introduced into the rectum to the height of the vesiculæ seminales, while the metallic brush faradizes the testicles and the entire surface of the penis, an erection may very often be produced during the treatment, as Onimus[1] remarks, and this fact has also been observed by myself.

Especially to be recommended is the application of

---

[1] Guide pratique d'Electrothérapie. Paris, 1882, p. 264.

weak induction currents during a longer time, because they are apt to revive the excitability of weakened nerves, as has been demonstrated by von Bezold and Engelmann.

Static electricity, *franklinization*,[1] or general electrization, is yet in rare use, although very good results have been obtained thereby in recent times. Eulenburg[2] asserts that carefully watched hydro-electric baths, electro-static air-baths, etc., are preferable to the process of general faradization and galvanization as indicated by Rockwell and Beard, because the latter are too complicated, cause loss of time, and are somewhat imperfect in their effect.

Hydro-electric baths[3] and general franklinization with the influence machine will be positively indicated in impotence dependent upon constitutional diseases, disturbances in the general nutrition, or states of weakness, but most of all in those forms brought about by general neurasthenia.

Next to the treatment of impotence by electricity comes the **local treatment** oftenest employed, which has again come into vogue.

In the first place comes *local cauterization*. Modern physicians have simply returned to Lallemand's method of treatment, which has been so much criticised. The principle has remained the same, the mode of application only having somewhat changed. I do not venture to say whether it is for the better, but prefer to discuss Lallemand's method briefly.

---

[1] Stein, Die allgemeine Elektrisation des menschl. Körpers. Halle, 1883.

[2] Sexuelle Neuropathie. Leipzig, 1895, p. 40.

[3] Eulenburg, Die hydroelektrischen Bäder. Wien und Leipzig, 1883, p. 76.

Lallemand cauterized the caput gallinaginis with a lapis-style or caustic-holder that formed part of a specially constructed instrument. To handle successfully Lallemand's instrument requires a certain dexterity which can be acquired only by practice; but when this dexterity is once acquired, the extent of the cauterization can be limited at will. Any one who has become skilled in Lallemand's method of cauterization will prefer it to all others, with the exception of the more simple method of cauterizing with the endoscope.

Every one will soon convince himself that such brilliant and prompt results as we read of in Lallemand's wonderful work cannot be obtained in our days by any method of cauterization that may be adopted. The method has remained the same, man is the same, and the nature of the disease is still the same as in Lallemand's time. The difference between our results and those of Lallemand can be explained by the fact that we are not the inventors of this method of treatment, and are therefore able to look upon our results with greater impartiality. It will enter no one's mind to accuse Lallemand of having intentionally distorted the facts; but every one can easily understand that Lallemand, who had correct ideas of the conditions in the urethra long before the endoscope came into existence, would see only the successful results of his method, and give little or no attention to his failures; also, that he overestimated the value of his invention. His brilliant success can be understood when we consider that in his time Lallemand was the only physician of renown who did not think it below his dignity to occupy himself with the different forms of sexual weakness; so that, as a natural consequence, people of the wealthy class suffering or imagining themselves to suffer from sexual weakness would flock to him from far and near. Comparing the number of psychically impotent with those that are affected with real impotence, we shall find that the former constitute the larger portion of those applying for relief. Now with these neurasthenics Lallemand's fame, combined with the great renown of his method, must have exercised a curative influence. Finally, we must not forget that great minds are not without a weakness or a hobby. I have a vivid recollection of a worthy clinician who died, unfortunately, too early. This gentleman thought he had discovered at one time a great epidemic of cerebro-spinal meningitis, and he became quite angry when his assistant merely sought for another explanation of the symptoms. When he had to treat obstinate patients who positively refused to show cutaneous hyperesthesia, he would pinch their skin, so as to produce for his own satisfaction utterances of pain which would indicate to him the existence of hyperesthesia.

Many German physicians took umbrage at the fact that they could not obtain the same brilliant results as Lallemand, and rejected his method altogether. In the eyes of other investigators Lallemand had presumed too much through his mania for writing. Although Lallemand has really treated the whole question too broadly, every one will prefer his prolixity to the scant treatment given to the subject by many authors of the present time. Of course, we no longer accept many of Lallemand's views; as, for instance, his exaggerated idea of the effect on the sexual power of riding, of tobacco, coffee, and tea. This, however, does not alter the fact that cauterizing after Lallemand's method does excellent service in pollutions, spermatorrhea, or impotence

TREATMENT.

caused by changes in certain portions of the urethra. The endoscope enables us now to obtain positive knowledge about this cause.

Special care must be taken in cauterizing, and it should be undertaken only by medical men who are perfectly familiar with the handling of the endoscope, and by them only under the control of this instrument. Cauterizing is thereby much facilitated, the need of a caustic-holder being dispensed with, since a lapis-style fastened on a piece of silver wire answers the purpose perfectly. The majority of physicians have for the endoscope a certain objection which, says Grünfeld, can exist only on the ground of their ignorance in regard to the details. This one application of the instrument should suffice to make its utility obvious to every thinking man.

The aëro-endoscope of Antal is excellent, but, I regret to say, cannot be used for cauterizing. Good service is obtained from an endoscope of Grünfeld, made of hard rubber and of sufficient length. At present I use only Nitze's endoscope, with its modifications by Oberländer and Kollmann. "The light is furnished by a platinum wire made incandescent by the electric current with which it is connected by conducting wires. The wire is introduced to near the visceral end of the tube and throws a strong light upon the exposed portion of the mucous membrane." This instrument is the best at our disposal, though I can only agree with Klotz[1] when he says that "Kollmann's instrument is by no means free from objectionable features and is certainly not as perfect as has been claimed."

---

[1] Klotz, The Practical Use of the Endoscope. Journal of Cutaneous and Genito-Urinary Diseases. New York, July, 1898, p. 313.

Immediately before the operation of cauterizing, the bladder should be completely emptied. One reason is to spare the patient the intense pain that would be induced by urinating soon after the operation, and another to prevent the risk of having the effect of cauterizing very much interfered with by an involuntary discharge of urine whilst the operation is going on. Nowadays it hardly seems necessary to mention any reasons, as no competent physician will introduce any kind of an instrument into the urethra without first having the bladder emptied and disinfecting the urethra itself.

The eye of the operator will guide him as to the extent of the cauterization, which, of course, will be determined by the pathological changes he observes, but in any case the cauterization must be confined within reasonable limits.

Lallemand[1] orders that after cauterization the patient is to bathe frequently during the first few days; he must cause regular defecation by clysters, and he must take light food, consisting of milk and vegetables. Again, he must avoid all bodily exertion and exposure to cold.

During the first two or three days following cauterization there is dysuria, and the urine contains a few drops of blood. In cases of particularly intense inflammation ice-bags may be applied to the perineum. If the pain is great, it may be relieved by the use of opium. A repetition of the use of the cautery cannot be thought of until all symptoms of inflammation have subsided. Lallemand thinks this will take a couple of weeks.

A similar although more feeble effect may be obtained by the use of *astringent injections*. I have already had

---

[1] Op. cit., tome iii. p. 401.

occasion to state that injections of cold water exert a certain stimulating influence, being frequently followed by erections. Astringents cause a stronger irritation, and the stimulation is in direct proportion to the strength of the injection.

Astringents for local excitation may be used also in the form of gelatin bougies. Dittel has had constructed a special porte-remede, by means of which he applies to the pars prostatica astringents in the form of small *urethral suppositories*.

Ultzmann has invented a urethral dropper for applying any astringent fluid or caustic to deeper portions of the urethra. A physician handling the endoscope can dispense with these instruments in the treatment of impotence if no special obstacle interfere with its introduction. Any difficulty in the introduction of the endoscope will also interfere with the introduction of the other instruments.

Zinc, alum, copper, or tannin may be used in the urethral injections. I especially recommend *tinctura ratanhiæ*, which can be used in solutions of different strength. 'One drop of pure tinctura ratanhiæ (Krameria triandra) on the pars prostatica has the same effect as one cauterization without any of the objectionable consequences that follow the latter procedure. Weak solutions of astringents can be injected with any ordinary urethral syringe. Stronger solutions had better be applied locally under the safe guidance of an endoscope.

Very good results are obtained in certain cases of impotence that accompany chronic gonorrhea and its complications by the use of *intravesical irrigations*. These can be performed thoroughly and easily with Valentine's intravesical irrigator.

The introduction of flexible *bougies* or *metal sounds* may be indicated in a great many cases, and was advised by Lallemand. Special results are obtained in hyperesthetic conditions of the urethra and the prostata. In such cases the introduction of a bougie or sound is accompanied by so much pain that the patient can with difficulty be induced to submit to a second introduction.

In cases of intense hyperesthesia it is best to begin with the insertion of flexible bougies as gently as possible. If the procedure is repeated daily, the hyperesthesia decreases markedly and gradually admits of the use of metal sounds. This would show that the treatment by sounds or bougies is of particularly good effect, especially in cases of precipitate ejaculation and of too frequent pollutions. The idea is probably quite correct that by the introduction of a heavy metal sound pressure and tension are produced in the pars prostatica, thus causing a stimulation and possibly an erection;[1] but this is of no particular value in the course of the treatment of impotence, because it soon loses its effect. In certain cases it may be useful as a temporary harmless excitant.

During recent years I have almost exclusively used metal sounds, because experience has taught me that after sufficient dexterity is acquired it is easier to insert a steel sound than a flexible bougie in cases of hyperesthesia of a high degree. When using a metal sound force is, of course, to be avoided, and the sound must be guided by the anatomy of the parts, and must be carried practically by its own weight into the bladder. In with-

---

[1] Ultzmann, Ueber Potentia generandi und Potentia cœundi. Wiener Klinik, 1885, Heft 1.

drawing the sound the free hand must take hold of the penis close to the meatus, and by gentle downward pressure protect the urethra from strain that might be induced by the escaping sound.

Many authors and most practitioners recommend *external applications* to the genitalia of various substances, such as tinctura nucis vomicæ, eau de Cologne, alcohol, etc. It is not to be denied that such applications increase for the time being the vascularity of the parts in question, and that the cutaneous nerves may thereby be for the moment excited. At any rate, the effect cannot be great, and therefore they may be ordered sometimes " ut aliquid fiat," especially in cases when the patient's uneasiness must be appeased. Concerning the tinctura nucis vomicæ recommended by Roubaud,[1] it would be interesting to ascertain whether some of it is absorbed through the skin of the glans.

We cannot approve of the application of *sinapisms* to the genitalia as recommended by Roubaud; because the least incautious management may do harm, and because no benefit can be derived from an erection produced by such a painful remedy.

We do not mean to waste more than a word here on *acupuncture* and *electropuncture*. They were formerly in use; were recommended by Lallemand, and in recent times by Roubaud; but no one nowadays would entertain the idea of indulging in such a procedure.

Local *surgical operation* must of course be adopted, when possible, for the removal of defects. The most frequently indicated operations are *circumcision* and the gradual *dilation of strictures*.

---

[1] Op. cit., p. 154.

The application of **massage**, general as well as local, and of **gymnastics**, may in many cases of impotence bring about very satisfactory results, especially when it is necessary to strengthen the body and to further assimilation.

General massage of the body, the Swedish movements, and our ordinary gymnastics can be used only with reasonable moderation. I do not think, however, that systematic exercise, even if carried to athleticism, can in any wise have an unfavorable influence on the sexual power, as is often stated. If some athletes are really disinclined to enjoyments in venery, the explanation is found in the fact that an occupation monopolizing time and causing bodily fatigue is not conducive to sexual pleasure. I have known some athletes who by no means despised sexual enjoyments. The idea that athletes are weak sexually probably arose from the fact that among the ancient Grecians all athletes were compelled to abstain from coition as much as possible.[1]

Substitutes for gymnastics proper are such other *bodily exercises* as riding, skating, etc.; and bicycling is to be recommended particularly. Lallemand[2] has said, "The action of the lower limbs has probably more of a direct influence upon the sexual organs." Of course, even these exercises must not be carried to the point of fatigue.

In some cases *traveling* may have marked effect. An interesting journey, and especially if made on foot, in part at least, is beneficial; it engages the mind and

---

[1] Busch, Allgemeine Orthopädie, Gymnastik und Massage. Ziemssen, Allg. Therap., Band ii. Theil 2, p. 20.

[2] Op. cit., tome iii. p. 384.

draws the patient away from bad company or from undesirable conditions. Many patients have returned perfectly cured from a proper journey. Voyages also, if not of excessive length, do very excellent service in proper cases, and especially in those where abstinence is positively indicated.

The *flagellations* that were known in ancient times,[1] and at one epoch enrolled in religious service, " la Discipline d'enhaut et la Discipline d'enbas,"[2] are in some sense a kind of massage. They are appropriated in our days by physically ruined debauchees as a means of stimulating the exhausted spinal cord. Ancient authors have a peculiar explanation for the stimulating influence of flagellations; for instance, Boileau:[3] "Cela pose, il faut de toute nécessité, que lors que les muscles lombaires sont frapez à coup de verges, ou de foüet, les esprits animaux soient repoussez avec violence vers l'os pubis, et qu'ils excitent des mouvements impudiques, à cause de la proximité des parties genitales : Ces impressions passent d'abord au cerveau, et y peignent de vives images des plaisirs défendus, qui fascinent l'esprit par leurs charmes trompeurs, et reduisent la chasteté aux derniers abois."

Our explanation of the effect of these flagellations differs considerably from the above. Besides, we must dispense with these means of treatment. In some cases we may apply along the spine different aromatic and irritative substances, which cause a local hyperemia and

---

[1] Roubaud, op. cit., p. 151.
[2] Histoire des Flagellans, traduite du Latin, de M. l'Abbé Boileau. Amsterdam, 1701, p. 5.
[3] Histoire des Flagellans, p. 307.

are good substitutes for the flagellations. We shall take no notice of urtication, moxa, vesicatories, and similar things.

Professional and other inventors have always endeavored to construct **apparatus and instruments** designed either to remove sexual impotence itself or to enable the impotent to introduce the non-erected or only partially erected penis into the vagina. I refer here to the paradoxical apparatus for the pretended enlargement of the penis by Roubaud, who has found no imitators; to the so-called "ventouse" by Mondat, mentioned also by Roubaud; and to numerous other devices that are almost useless.

The first step in a practical direction was made by an inventor unknown to me, who constructed a small instrument in which a flourishing trade was carried on for a while, and which I described in the first edition of my "Pathology and Therapy of Sexual Impotence." There existed at that time also medical moralists who declaimed against such instruments; but, in spite of this, I was confirmed in my opinion, expressed at that time, that even a conscientious physician in cases especially worthy of regard, and when every other remedy had failed, could take upon himself the responsibility of advising the use of such an instrument. Shortly after the appearance of my work, letters from every part of the world came to me from physicians inquiring concerning the instrument. I was not in a position to advise them as to where to obtain it, as the one I had came to me through a patient.

The instruments were made of German-silver, silver, or gold, and consisted of two delicate splints connected at the base by a metal ring, and at the upper end by a soft-rubber ring. In the jargon of elderly bon-vivants it used to be called "the sledge," and fairly fulfilled its purpose, when made exactly to measure, in spite of considerable inconvenience arising from the fact that the penis is not supported at the base, and that the instrument certainly cannot remain unnoticed and unfelt by the female partner.

The various chains, belts, plates, and other contrivances which, as Fürbringer says, bring to mind the amulets of old, we shall not consider at all, as they are calculated solely to abuse the ignorance and credulity of the masses, and are of greater utility to the rascally vender than to the gullible purchaser. A real progress in the direction of the mechanical treatment of sexual impotence is shown by Paul Gassen's devices. This "erector" consists of a doubly coiled spiral provided at both ends with knob-like masses. The instrument is twisted round the member in such a manner that the greater button-like extremity is placed in the region of the anus, the smaller one on the right side of the frenulum. The first turn will consequently be placed on the dorsal side of the bases of the penis, enabling it to exercise pressure on the vena dorsalis, thus checking the reflux of blood from the corpora cavernosa.

The interest of European physicians for these instruments was awakened through Krafft-Ebing's [1] expert opinion, as well as Fürbringer's [2] article in the "Zeitschrift

---

[1] Gerichtliches Gedachten über ein von dem Techniker Paul Gassen erfundenes Instrument zur Behebung der Impotenz, genannt Erector. Friedreich's Blätter für gerichtliche Medicin und Sanitätspolizei, 1897, Heft 3, p. 217.

[2] Zur diätetischen und physikalischen Behandlung der Impotence. Zeitschrift für diätetische und physikalische Therapie, 1898, Band I. Heft 1.

für diätetische und physikalische Therapie," and numerous experiments on the subject followed.

I shall first give the views of these and some other authorities on the subject, and then the results of my own experience.

Krafft-Ebing, in his well-known opinion, given as an expert before the royal court of justice of Cologne, said, among other things: "Paul Gassen's erector is in general adapted to afford the results claimed for it in the circular, in spite of this being too full of self-praise, in so far as it promotes the erection, and gives to the penis at least part of the rigidity requisite for the inimissio in vaginam." Further: "Conditions of absolute impotence are, however, rare, and are caused only by severe vertebral and nervous diseases. In medical practice we have, in a vast majority of cases, to do merely with relative impotence through physical causes (exhaustion as a consequence of excesses of individuals who have abused the natural sexual pleasures or in consequence of onanism) or psychical (imaginary obstacles, fear of failure, etc.). Here a considerable or virtually the full power has been preserved, and the erector may, in the first case, compensate for the failing remainder of power, and afford, as it were, a crutch for the lame; in the latter case it acts, in combination with its mechanical action, psychically, and, awakening the confidence in the required capacity, it compensates for the imaginary obstacles called forth by the mistrust of his own power, under circumstances preventing erection; just as, for example, any one suffering from agoraphobia, being in this psychical anomaly incapable of crossing a square, is enabled to do so when accompanied at starting merely by a child.

"The inventor of the erector, in his circular, had in view as a layman merely the mechanical effect of his instrument, and those debilitated and enervated through sexual abuse. He had no idea that there are very many sufferers with impotence who are so from no fault of theirs, through psychical influence, and only requiring psychical aid.

"In so far, however, as the instrument is adapted, in cases of merely relative impotence, to accomplish important ends, at least to facilitate the sexual act mechanically, it gives a quasi-guaranty of the result in the case of the psychically impotent, and frees him, through the success attained, of his psychical obstacles, and thus renders him absolutely able. In this respect we might even speak of a (psychical) cure through the erector.

"As the potentia cœundi is a necessary condition for the potentia procreandi, the erector appears eventually also adapted for insuring the latter capacity.

"As, however, the instrument essentially facilitates the conditions for the accomplishment of coitus, impotence being for the one who suffers from it a physical and psychical evil, its use, rendering possible, as it does, the accomplishment of a function natural and important both for the body and for the soul (psyche), can in general operate only favorably, except in case one is misled by the artificial contrivance to excesses in coitus—a circumstance, however, which should not be attributed to the instrument, but to the wearer.

"In view of the very great frequency of impotence in modern society, and the significance of this evil for those who suffer from it, as well as the imperfection of medicinal and physical remedies, the fact is readily understood that mechanical expedients have long ago been devised by physicians to come to the aid of failing or weakened power." Further: "Paul Gassen has, as we see, scientific medical predecessors in the domain of invention of mechanical contrivances for the removal of impotence, and the need of such will always exist, since, on the one hand, the medicaments at our disposal are only exceptionally able to cure impotence, and, on the other hand, the most important interests of patients are at stake: health, fitness for marriage, capacity for procreation, and, in the negative case, bodily and mental disease, suicide, adultery, etc.

"Accordingly, Paul Gassen's erector appears at the present time as the best expedient for the improvement and attainment of sexual capacity for all who are in the sad condition of needing the service of such mechanical contrivances, and medical science would have been under obligation to the inventor if he had placed his invention at the disposal of its representatives for trial and application, instead of making it from the start an object of advertisement and mercantile enterprise."

Fürbringer also thinks that Gassen's erector must be considered an instrument constructed on rational principles, when used under proper circumstances. In a simple and ingenious manner it has completed the principle of the "sledge," under the form of an elastically flexible serpentine coil in such a manner that opposite to each point of pressure there is a corresponding point without such pressure. Prevention of natural erection in any way is accordingly excluded. Fürbringer joins also in the view of the two medical experts who, together with Krafft-Ebing, expressed themselves in those lawsuits, that by means of the erector, which, for the rest does not always fit equally well, the erector may be maintained after ejaculation.

My own experiences with Gassen's apparatus are, with respect to the erector, decidedly in harmony with the views just quoted. The erector, when made according to correct measurement, supports the flaccid member so that it can be introduced into the vagina. This is particularly easy of accomplishment by partial or even only slight erection, and through the movements of coitus thereby rendered possible the erection is augmented or even completed in cases where this is possible at all, and a gradual vanishing of the erection, as usually happens in so many cases of neurasthenic impotence, is prevented.

Gassen's compressors are no new idea; similar expedients were known to me years ago, but his compressors are preferable to the formerly used annular compressor, because they are easily put on and quickly removed. The compressors are hardly capable by themselves of causing an erection, but they can maintain and increase an already existing partial filling of the cavernous tissue, exerting, as they do, the necessary pressure to check the reflux of the venous blood, but not compressing the arteries.

The suction-pump called by Gassen the "cumulator" can, in particular cases, be used for a kind of gymnastics of the erectile tissue. It is new in execution, but the principle is old.

Considering the experience I have had with "the sledge," we must assume that the erector will bring aid in many cases of impotence. Should any one abuse the apparatus in practising excesses, this would be the fault, as Fürbringer correctly remarks, of the individual, and not of the apparatus. When I remember the case communicated by me at the time, in which a psychically impotent young man was able to accomplish copulation only when he had the so-called "sledge" as a surety with him, without, however, ever really making use of it, it is to be expected that neurasthenic impotence will furnish a good field for these mechanical remedies.

The use of Gassen's "ultima" will hardly ever be advised by any physician, though extreme and extraordinary cases may justify even this kind of a last refuge.

In the treatment of most forms of impotence special weight must be attached to the **regulation of the sexual life.** Here also we must carefully consider the form of the disease, as well as the constitution and general disposition of the patient. There may be cases in which the physician is compelled to order absolute continence, in accordance with what we have previously stated on this subject. This advice will generally be indicated only in those cases in which the patient is to be convinced that he is in reality not impotent.

Generally, however, the physician will find it necessary to order the patient to have regular intercourse and this advice may have to be given sometimes from the very beginning, sometimes later in the course of the treatment. The medical man must feel it incumbent upon himself to do his duty conscientiously and according to his best knowledge, even though he may fail to sustain his "dignity." The times have gone by when the doctor walked along gravely adorned with his doctorate's hat, his stick, and his periwig; they have gone by together with the periwig and the pigtail. It is well that they are passed, and it is to be hoped that they are forgotten. In our days we wish to advise and help the sick. Our dignity does not suffer if we order regular sexual indulgence.

I notice with very much satisfaction that I am not alone in my ideas upon this subject, the expression of which, about nine years ago, earned for me so much adverse criticism. Prince A. Morrow,[1] for instance, says, "The exercise of the sexual organs within certain bounds undoubtedly has the effect of strengthening, invigorating, and preserving them in their full integrity."

Such a prescription may be easily given, but for its compounding the patient cannot be referred to a druggist. The physician is not supposed to give himself trouble about the procuring of a suitable companion; but even on this question the doctor may not refuse advice, especially when he is rendered competent by knowledge of medicine and the world.

In these cases physicians have, in all times, advised *marriage*. Now, marriage is a delicate and important affair. Still, I do not entertain the views of those authors who will not admit that matrimony should be advised from the therapeutic standpoint. I believe that matrimony is a contract in which one party agrees to give something for something to be obtained in return. It is a serious matter to advise matrimony, for no physician wishes to carry on his conscience the misfortune that may arise from another man's matrimonial venture. It must, however, be admitted that marriage with a suitable person is the safest and most reliable remedy for many forms of impotence, and it serves also as a great preventive against the contraction of impotence. On the other hand, most patients object to matrimony. Many even have an aversion for marriage in general, and think they have sufficient reason for their belief. In such a case the phy-

---

[1] A System of Genito-Urinary Diseases, vol. i. p. 1003.

sician will, of course, not insist on his advice. Others, again, dare not venture to enter into matrimony because they believe themselves to be unable to fulfil their conjugal duties. If the physician finds this to be true, there is again every reason for not advising matrimony.

A man who is even moderately virile had better be dissuaded from marriage, because his sexual weakness would render his hymeneal happiness doubtful, unless he should happen to find a wife not particularly given to sexual pleasure. Here very great discernment and caution should prevail, because such a quality is not to be read on the forehead or face of a woman. See here the beautiful diplomatic language with which Rosenthal [1] expresses his opinion: "In case of recovery or lasting improvement in the sexual power, a subsequent marriage to a person of calm temperament may be allowed in order to preserve the restored condition."

As marriage is a heroic and very dangerous remedy not accessible to every one, and as a mistake in this affair is so difficult to correct, many a convalescent patient will be compelled to have recourse to other connections than hymeneal in order to satisfy his sexual desire, if he does not want to become impotent again or to be troubled again by morbid pollutions. He must satisfy this natural want regularly, and the act cannot be called immoral simply because it is accomplished out of wedlock. Such connections may be unknown to dried-up pedants who have never been young, but every man gifted with a heart and physical power is familiar with these liaisons in all their variations.

---

[1] Ueber den Einfluss von Nervenkrankheiten auf Zeugung und Sterilität. Wiener Klinik, 1880, Heft 5, p. 141.

An individual weak in sexualibus is seldom lucky enough to captivate a woman's heart, and prostitutes are his only recourse. Modern prostitution is, as Mantegazza[1] expresses it, "The product of the Christian virtue, which wants a perfect man, and the animal instinct, which drives the man into a woman's arms."

We must consider that the functional capacity in intercourse with a prostitute is not in proportion to the actual sexual power, and that real prostitutes—*i.e.*, those "whom any one can buy"[2]—are too repulsive to please the good taste of some patients. Again, we must consider the dangers of venereal diseases. Pamphlets bearing the pompous title of "Prevention of sexual contagion"[3] certainly prevent nothing whatever, and, moreover, tell only facts known to every one beforehand.

It is obvious that a physician can in these matters give little advice to his patient, but must entrust him to good luck. Besides, when the physician has done his duty toward an individual and cured him completely or partially, he can generally yield him to his fate with tranquillity, because most of the sexually weak persons who take medical advice are of the intelligent and wealthy class of society.

The sanguine hopes that were entertained once about the result of *inhalations of oxygen* have proved, by thorough investigation and experiments, to be a mere fata Morgana. Sometimes, however, this remedy may be given a trial, because it may be beneficial, and especially in cases complicated with anemia, leukemia, diabetes, dyspepsia, and chronic weakness.

---

[1] Gli amori degli uomini. Milano, 1886, vol. ii. p. 201.
[2] Jeannel, De la prostitution. Paris, 1874, p. 190.
[3] For instance, Dr. A. Theod. Stamm, Zürich, 1886.

The therapeutics of impotence has been greatly enriched by Motschutkovsky, who, whilst applying Sayre's corset, noticed that the body lengthened during **suspension**. Repeated measurements proved that this lengthening is due to the stretching of the vertebral column, and varies between two and one-half and five centimeters. This led Motschutkovsky to the idea of trying suspension in diseases of the spinal cord. The result was excellent: several troublesome symptoms vanished during the treatment. Of chief interest for us in this experimentation is the circumstance that in every case a certain number of suspensions removed all the pre-existing disturbances in the sexual feelings and power. In Charcot's clinique at Paris the same fact was established, along with decided improvement in functional diseases of the bladder, so often accompanying tabes; and also the cure of neurasthenic impotence. Bernhard reports a case in which, after nineteen suspensions, erection and pollutions reappeared after having been absent for over a year.

I have treated with suspension numerous patients threatened with paralytic impotence, and in several cases obtained considerable improvement. After a few suspensions erections and libido partially returned. I never noticed lasting results, however. Everything returned to the old state a few days after the cessation of the suspensions, and, moreover, the suspensions themselves lost their efficacy after a few weeks. These suspensions in paralytic impotence may be compared to one of the last lashes given to a jaded beast of burden.

Since, however, suspensions have had some effect on paralytic impotence, a disease in which therapeutics is usually powerless, it was certainly to be expected that they would produce better, or even satisfactory, results in forms of impotence that are easier to cure. Indeed, I have obtained very favorable results, and sometimes a perfect cure, in various cases of sexual neurasthenia.

One of these cases, which I published at the time, I shall reproduce here. B. K., a lawyer, thirty-four years old, somewhat thin, though always healthy, had in his youth passed through his experiences with onanism. After that he was on intimate terms for six years with a woman two years his senior. During this entire connection he never experienced any signs of impotence, for he was all the time able to accomplish the act to his heart's content. In the rare intercourse he had with prostitutes during those six years he had a few failures in coition, but did not attach any importance to them. For five weeks before he applied to me he had vainly endeavored to accomplish coition, although his opportunities had been most favorable and the responsiveness of his companion all that could be desired. Hence he was in a state of great excitement and believed himself to be quite and permanently impotent. After a fruitless attempt with nux vomica I proceeded to the use of suspensions, after having represented them to be a sure and infallible curative remedy. I was informed by the patient that he had erections in the night following the first suspension, and after the fifth he accomplished coition without my permission, and he assured me he did not need further medical treatment. After the ninth suspension I dismissed him from my care.

Although the result was satisfactory in this case as well as in many others, I was not convinced that these suspensions had an aphrodisiac power, because I knew that psychically impotent persons are sometimes cured by the most inefficient means. To test the suspensions a little further I frequently used them on healthy persons. Though imagination may have had more or less influence on these persons, they asserted that the suspensions had a stimulating effect on the sexual desire.

I tried suspensions in a case of frigidity on the part of a married lady. Both she and her husband attributed the absence of children to this cause, but an examination was refused. Although in this case suspensions had no effect whatever, I do not contend that they have no efficacy in frigidity, which is not infrequent in women, and I think further experiments advisable. In the case just mentioned, I believe the wife had an aversion for her husband, although she assured me to the contrary.

For these suspensions I invariably use Sayre's apparatus modified by Motschutkovsky. The act of suspending is accomplished gradually and with great care. The first suspension never lasts over two minutes, but in most cases I have to limit it to one minute or less. By degrees, as the patients feel less afraid, the duration can be prolonged; but I have never gone beyond five minutes, not even when enthusiastic patients requested it. Almost without exception they were made every second day.

As to the manner in which the suspensions act, we must confine ourselves more or less to conjecture. Is the cause the momentary change brought about in the position of the spinal cord and its nerve-trunks? Is it the tension of the more peripherally situated nerves? Is it the increased pressure of the blood and the increased circulation of blood in the vessels of the spinal cord, which possibly is caused by the increased blood-pressure?

Of course, suspension is out of the question with persons suffering from defects of the heart, atheroma, aneurism, emphysema, cavities in the lungs, inclination to hemorrhage from the lungs, epilepsy, apoplexy, also anemia of a high degree. Every physician should devote his attention to these evils before commencing any treatment for impotence.

Finally, I wish to mention a therapeutic procedure that has enthusiastic supporters, but also many bitter opponents. I mean **hypnotism**, the therapeutic measure by which real miracles are sometimes accomplished, in spite of statements to the contrary.

I have personally obtained very good results in several cases that were suitable for such treatment; but at the same time one should be guarded against self-delusion. In the treatment of psychical impotence suggestion is, indeed, the means without which we can expect but little result; and from suggestion to hypnotism there is only one step.

Bernheim[1] was able to influence menstruation by hypnotic suggestion. Krafft-Ebing[2] wrote thus: "The task of posthypnotic suggestion is in such cases to suggest a dissuasion from the impulse to commit masturbation; to create, by suggestion, a feeling against homosexual desires; to induce a consciousness of virility, and to arouse heterosexual desires."

Baron v. Schrenck-Notzing relates one of his cases that beautifully illustrates the therapeutic power of suggestion.[3] Another case may be mentioned, that of Tessie

---

[1] De la suggestion. Paris, 1888, pp. 557–563.
[2] Psychopathia sexualis. Stuttgart, 1890, p. 225.
[3] Ueber Suggestionstherapie bei conträrer Sexualempfindung. Internat. klin. Rundschau. Wien, 1891, No. 26.

quoted by Casper.[1] Many cases of neurasthenia have been cured by hypnotic suggestion. Bérillon reports more than twenty-two cases of nocturnal incontinence of urine, and four cases of irresistible onanism in children.[2]

When all of this is considered, we may justly expect hypnotic suggestion to prove to be a powerful remedy in onanism, morbid pollutions, and various forms of impotence.

In conclusion, I must speak of **organotherapy**, which, though still in its inceptive stadium, promises great results. My personal experience with Brown-Séquard's liquor testiculorum is best expressed in the words that Eulenburg[3] uses when he speaks of his own experience with Poehl's spermine: viz., "The results are varying and unequal, though sometimes surprisingly favorable without any disagreeable collateral action." G. Hirsch[4] is inclined to think that Brown-Séquard's testicular emulsion contains both substances, which when ejected are useful, and substances which have a disturbing action on the metabolism. Professor Poehl was able to isolate a substance, spermin, which he thinks is contained in all organs used therapeutically. Hirsch has tried Poehl's spermin in cases of anemia, tabes dorsalis, and endarteritis obliterans, and by his own observations communi-

---

[1] Casper, Impotentia et Sterilitas virilis. München, 1890, p. 99.
[2] Bérillon, Ueber die Indicationen für die hypnotische Suggestion in der Psychiatrie und Neuropathologie. Vortrag, gehalten in der Section für Neurolgie und Psychiatrie des X. internat. med. Congresses zu Berlin. Ref. d. Wiener med. Presse, 1891, No. 3.
[3] Sexuale Neuropathie. Leipzig, 1895, p. 41.
[4] St. Petersburger med. Wochenschrift, 1897, No. 7; British Medical Journal, 1898.

cated to him he has arrived at the following conclusions. The spermin has no specific action at all in particular diseases. It seems, however, to control in some way the metabolism or intraorganic oxidation, and by the removal of accumulated waste products to disencumber the nervous system, and so finally to favor the vis medicatrix naturæ.

On the whole, a trial may be ventured, and further experiments are advisable. Brown-Séquard's fluid of the best quality has always been sent to me directly by a druggist in Geneva.

CHAPTER X.

SPECIAL THERAPEUTICS.

In this section I shall briefly discuss the curative methods as they are indicated in the various forms and grades of impotence.

The therapeutics of congenital and acquired malformations and defects in the sexual organs will be indicated in each case by an examination, and if anything is to be done at all, some surgical operation will have to be performed in nearly all cases.

Impotence that has come in the track of different pathological conditions can be treated only after removal of these causal conditions, this being accomplished by well known and approved methods. If impotence should still remain after the removal of the causes, we must then call into service all the stimulating means at our disposal, together with a selected diet appropriate for convalescent patients. The means indicated then are, first, hydro-therapeutic measures, electricity, river-bathing or sea-bathing, and eventually the internal use of nux vomica and oxygen-inhalations.

The treatment of inherited sexual weakness is very difficult. First of all the sexual desire must be awakened. This cannot be the business of the medical practitioner, but must be left to friends or relatives of the patient, who may be advised by the physician. After the sexual desire has once been awakened, we may proceed to the use of various means of stimulation in order to arouse the sexual power that possibly lies dormant; but experience teaches that this is hardly ever successful. Fortunately, sexual weakness of a high degree is seldom inherited.

The medical science is of but little use in the cure of perverse sexual sensation. However, education, or perhaps hypnotic suggestion, may be of some benefit.

When sexual neurasthenia and impotence have been induced by bad management of the sexual power, therapeutics must choose various means in accordance with the state and physical strength of the patient. In every case the treatment must begin with the regulation of the sexual life, and in some isolated cases it may be advisable to order continence of moderate duration. It will be proper to order for individuals in a declining condition, besides a correct diet and certain medicaments, gymnastics, massage, hydro-therapeutics, oxygen-inhalation, or general electrization of the body. In patients who are still physically strong you will find indicated, besides ordinary food, which need not be too delicate, hydro-therapeutics, local electricity, the sound or bougie treatment, local injections, cauterization of the caput gallinaginis, and, possibly, suspensions.

As above mentioned, in case onanism or spermatorrhea exists, we must endeavor energetically to correct them.

In the case of impotence originating from continence, which, however, is exceedingly rare, it is the physician's duty to reawaken the dormant virile power; that is, he must stimulate vigorously the sexual nerve-centers and tracts that have grown indolent from want of proper excitation. In this case, bad company, which has the reputation of spoiling good manners, may do some good. If necessary, recourse may be had to electricity, douches, suspensions, and local stimulation of the mucous membrane of the urethra, especially of the caput gallinaginis.

Most of all, purely neurasthenic impotence presents numerous difficulties, because an unwise treatment may easily do harm. On the other hand, a physician who knows how to think will find this a most satisfactory field for his action, because good results are here more easily obtained than in any other form of impotence. A wise, psychical treatment does much in such cases, but cannot constitute the entire treatment. Neurasthenic conditions cannot be cured by simply telling the patient that he is not afflicted. Often hypnotic suggestion does very good service.

In so-called irritable weakness we may exercise some good influence against precipitate ejaculation by toning up the body in general and the sexual power in particular. The means suitable for that effect are hydrotherapeutics, sea-bathing, river-bathing, gymnastics, massage, and the psychrophor and the sound to lessen the sensibility of the mucous membrane of the urethra. Every exciting means must be avoided in case the irritable weakness is attended by sexual agitation in a high degree.

Edward Martin[1] recommends for the avoidance of premature ejaculation, " where the patient is educated and has a trained mind, the concentration of the latter upon some act of memory, such as recalling a recitation, or upon some calculation in mathematics." I am afraid

---

[1] Impotence and Sterility. Hare: System of Practical Therapeutics. Philadelphia, 1892, vol. iii. p. 665.

that such a scheme would work in but very few cases, because we know quite well that anything which distracts the mind from the sexual act is apt to impair the necessary erection.

I have found that one or two glasses of good beer taken before the act very often controls precipitate ejaculation.

In other forms of sexual neurasthenia the treatment will be determined in each case by such causes as can be discovered, and the case may have to be treated symptomatically. In all cases the frequency of intercourse must be regulated.

Impotence induced by occupation may also be a subject for treatment, but good results can then be obtained only when one succeeds in annihilating, or at least diminishing, the injurious influences of the occupation.

Senile impotence can never be the subject of rational medical treatment, though one may sometimes sympathize with an amorous old man.

INDEX.

Ablutions, 246.
Absence of the penis, 87.
Acupuncture, 261.
Acute diseases and sexual vigor, 97.
Age in etiology of impotence, 81.
Albuginea of the corpora cavernosa, 43.
Alcohol, influence on virility, 113.
Ampulla of Henle, 35.
Anæsthesia sexualis, 126.
Anatomy of male genitals, 30.
Anemia and impotence, 100.
   from onanism, 171.
Animals, perverse sexual feeling for, 133.
Antipyrin, effect on virility, 121.
   in spermatorrhea, 234.
Aphrodisiac foods, 237.
Apparatus for impotent men, 264.
Arsenic, effect on virility, 119.
Artists, sexual power in, 203.
Astringent injections, 258.
Atrophy of testicles from onanism, 174.
   from sexual excess, 159.
Atropin in spermatorrhea, 234.
Azperger's rectal-cooling sound, 250.

Balneological treatment, 249.
Bathing, in spermatorrhea, 236.
Baths, 244.
Beer, effect on virility, 114.

Belladonna, effect on virility, 121.
Bladder disease and virility, 110.
   rheophore, 252.
Bodily structure in etiology of impotence, 80.
Body, effect of impotence on, 25.
Bougies, 260.
Brain disease and virility, 102.
Bromids in spermatorrhea, 234.
Brucin, 241.

Camphor, effect on virility, 121.
   in spermatorrhea, 234.
Cancer of the testicles and virility, 112.
Cannabis indica, effect on virility, 118.
Cantharides, 239.
Cantharidin, 239.
Capon-obesity, 21.
Caput gallinaginis, 39.
   cauterization of, 255.
Carbon-douche, 250.
Castoreum, 242.
Cauterization, 254.
Cavernous portion of the urethra, 41.
Character, effect of impotence on, 22.
Chlorosis and impotence, 100.
Chronic diseases and sexual power, 99.
Cocain in treatment of impotence, 243.
Coffee, effect on virility, 116.

INDEX.

Coition, after-effects of, 148.
  during menstruation, 221.
  excess in, 141.
  frequency of, 141, 219.
  frequent, effect on semen, 72.
  immoderate, consequences of, 152.
  physiology of, 45.
  position in, 220.
  process of, 56.
  time for, 220.
Coitus. See *Coition.*
  interruptus, 69.
Cold and impotence, 100.
  applications in spermatorrhea, 236.
Colliculus seminalis, 39.
  diseases of, and virility, 111.
Congenital impotence, 122.
  diagnosis of, 211.
  prognosis of, 214.
  special therapeutics of, 280.
  malformations and defects of the sexual organs, 87.
Consecutive impotence, 97.
  diagnosis of, 211.
  special therapeutics of, 280.
Constipation as a cause of pollutions, 180.
Continence, excessive, effect of, 188.
Copulation. See *Coition.*
Cornutinum citricum in spermatorrhea, 235.
Corpora cavernosa, 42.
  changes in, 96.
Corpses, defilement of, 135.
Corpus Highmori, 33.
Cowper's glands, 41.
  secretion of, and semen, 68.
Crista urethralis, 39.
Cryptorchidia, 92.
Cumulator, 269.

Damiana, 244.
Dartos, 31.
Death from sexual excess, 159.
Diabetes and virility, 99.
Diagnosis of impotence, 209.

Diet in treatment of impotence, 237.
Digitalis, effect on virility, 117.
Diphtheritis and impotence, 98.
Douche-baths, 247.
Dry cold, 250.
  warmth, 250.
Ductus ejaculatorius, 36.

Ejaculation, 59.
  center for, 60.
  precipitate, 192.
  premature, treatment of, 282.
Ejaculatory ducts, 38.
Electricity in spermatorrhea, 234.
Electropuncture, 261.
Electrotherapeutics, 251.
Endoscope, urethral, 257.
Endoscopic examination, importance of, 209.
  in pollutions, 184.
Epicures, sexual power in, 204.
Epididymis, 34.
Epispadia, 91.
Erectile tissue, defective development of, 89.
Erection, physiology of, 51.
Erector, 265.
Ergotin, 242.
Etiology of impotence, 79.
Excess in venery, 137.
Exercises in treatment of impotence, 262.
External applications, 261.

Faradic current, 253.
Flagellations, 263.
Flapping, 246.
Flogging and onanism, 169.
Food in treatment of impotence, 237.
Foods and virility, 117.
  stimulating, and onanism, 170.
Forms of impotence, 87.
Fowler's solution in spermatorrhea, 234.
Franklinization, 254.

INDEX. 287

Frauding, 69, 222.
Frenulum, malformation of, 90.
Friction, 246.
Frigidity, 126.
  from onanism, 173.
  from sexual excess, 153.

Galvanic current, 252.
General diseases as a cause of pollutions, 181.
Genitals, acquired defects of, 93.
  appearance of, 79.
  congenital malformations and defects of, 87.
Glands of Littré, 41.
Gonorrhea and virility, 109.
Gymnastics, 262.
  in spermatorrhea, 233.
  sexual, 146.

Half-baths, 248.
Heredity in etiology of impotence, 81.
Hermaphrodites, 93.
Hydatis Morgagni, 34.
Hydrocele and impotence, 95.
Hydro-electric baths, 254.
Hydrotherapeutics in spermatorrhea, 232.
Hydrotherapy, 244.
Hygienic living in treatment of impotence, 236.
Hyperesthesia of sexual organs from onanism, 173.
Hyperesthesis of sexual sensation, 130.
Hypnotism, 277.
Hypochondriac impotence, 195.
Hypospadia, 91.

Idiocy, impotence in, 135.
Idleness and onanism, 170.
Impotence and matrimony, 22.
  classification of varieties of, 84.
  congenital, 122.
    diagnosis, 211.
    prognosis of, 214.
    special therapeutics of, 280.

Impotence, consecutive, 97.
  diagnosis of, 211.
  special therapeutics of, 280.
definition of, 28.
diagnosis of, 209.
effect on the body, 25.
effect on character, 22.
effect on the mind, 20.
etiology of, 79.
forms of, 87.
frequency of, 26.
from acquired defects of the genitals, 93.
from congenital malformations of the genitals, 87.
hypochondriac, 195.
inherited predisposition to, 122.
misfortune of, 20.
neurasthenic, 136.
  diagnosis of, 211.
  prognosis of, 214.
  special therapeutics of, 282.
paralytic, 156.
professional, 202.
prognosis of, 214.
prophylaxis of, 217.
psychical, 195.
  diagnosis of, 213.
  treatment of, 228.
relative, 199.
senile, 205.
temporary, 198.
treatment of, 226.
Impotent man, appearance of, 21.
Incontinence of urine and sexual weakness, 124.
Indian hemp in treatment of impotence, 242.
Inguinal hernia and impotence, 95.
Inhalations of oxygen, 273.
Inherited predisposition to impotence, 122.
Injections, astringent, 258.
  urethral, 250.
Insanity and impotence, 102.

Instruments for the impotent, 264.
Intravesical irrigations, 259.
Iodin, effect on virility, 120.
Iron, 242.
Irritable weakness, 192.
　special therapeutics of, 282.
Isthmus urethræ, 40.

Krafft-Ebing's classification of impotence, 84.

Lacunæ Morgagni, 41.
Lead-poisoning and virility, 120.
Libido sexualis, 47.
Liquor testiculorum, 278.
Littré's glands, 41.
Local treatment, 254.
Love-potions, 239.
Lupulin, effect on virility, 121.
Lust-murder, 130.

Male genitalia, anatomy of, 30.
Marriage in treatment of impotence, 271.
Massage, 262.
Masturbation. See *Onanism*.
　causes of, 164.
　excessive, 160.
Medicaments and virility, 113.
　in treatment of impotence, 239.
Menstruation, coition during, 221.
Menthol, effect on virility, 122.
Mercury, effect on virility, 120.
Metal sounds, 260.
Mind, influence of impotence on, 20.
Mineral waters in spermatorrhea, 283.
Monobromated camphor, effect on virility, 121.
　in spermatorrhea, 234.
Monorchidia, 92.
Morphin, effect on virility, 118.
Musculus cremaster, 31.
Musk, 242.

Narcotics in impotence, 242.
Narrowness of the orificium externum urethræ, 89.
Nerve-disease and impotence, 101.
Nervousness and virility, 127.
Neurasthenia and impotence, 136.
　and virility, 103.
　from onanism, 172.
　from sexual excess, 153.
Neurasthenic impotence, 136.
　diagnosis of, 211.
　prognosis of, 214.
　special therapeutics of, 282.
Niter, effect on virility, 121.
Nitze's endoscope, 257.
Nux vomica, 241.

Obesity and virile power, 99.
Obstipation as a cause of pollutions, 180.
Old age, sexual power in, 205.
Onanism, causes of, 164.
　consequences of, 171.
　cutaneous diseases and, 170.
　excessive, 160.
　flogging and, 169.
　idleness and, 170.
　lascivious reading and pictures and, 169.
　prophylaxis of, 218.
　riding and, 170.
　sedentary life and, 169.
　seduction and, 166.
　sewing-machines and, 171.
　stimulating foods and, 170.
　tight garments and, 171.
　uncleanliness and, 170.
　unwise parents and, 166.
　why it causes pollutions, 181.
Opium, 242.
　effect on virility, 118.
Organotherapy, 278.
Orgasm, sexual, 47.
Oxygen, inhalations of, 273.

INDEX.

PARALYTIC impotence, 156.
Pars membranacea urethræ, 40.
    prostatica, 39.
Pederasty, 132.
Pendulous abdomen and impotence, 100.
Penis, absence of, 87.
    curvature of, 89.
    excessive development of, 88.
    flabbiness of, 89.
    loss of, 93.
    smallness of, 88.
Penis-bones, 97.
Perverse sexual feeling, 128.
    for animals, 133.
    for inanimate objects, 134.
    for the other sex, 130.
    for persons of the same sex, 132.
Phenol, effect on virility, 122.
Philters, 239.
Phimosis, 90.
Phosphorus, 240.
Phthisis and sexual vigor, 98.
Physiology of the sexual act, 45.
Poehl's spermin, 278.
Poisons and virility, 113.
Pollutions, 174.
    during sleep, 176.
    in the waking state, 177.
    morbid appearance of genitals in, 185.
    diagnosis of, 183.
    state of health in, 185.
    treatment of, 230.
Precipitate ejaculations, 192.
Premature ejaculation, treatment of, 282.
Prepuce, absence of, 89.
    superfluity of, 90.
Professional impotence, 202.
Prognosis of impotence, 214.
Prophylaxis of impotence, 217.
Prostate, 37.
Prostatic disease and virility, 109.
    secretion, 38.
    and the semen, 67.

Prostitutes in treatment of impotence, 273.
Psychical impotence, 195.
    diagnosis of, 213.
    treatment of impotence, 228.
Psychrophor, 249.
Puberty, 45.

QUININ, 242.
    effect on virility, 122.

RAIN-BATHS, 247.
Relative impotence, 199.
Resorption of semen, 63.
Rest in treatment of impotence, 239.
Rete vasculosum testis, 33.
Rheophore, bladder, 252.
Riding and onanism, 170.
    effect on sexual power, 205.
River-bathing, 248.
Rubbing down, 246.

SALICYLIC acid, effect on virility, 120.
Satiety for sexual pleasures, 154.
Scholars, sexual power in, 203.
Scincus marinus, 243.
Scrotum, 30.
Sea-bathing, 248.
    in spermatorrhea, 233.
Seasons, influence on sexual power, 83.
Secale cornutum, 242.
    in spermatorrhea, 234.
Sedentary life and onanism, 169.
Seduction and onanism, 166.
Semen, 61.
    appearance of, 70.
    chemical analysis of, 70.
    effect of frequent coition on, 72.
    constitution of, 66.
    effect of loss of, 150.
    production of, 61.
    quantity ejaculated, 69.
    resorption of, 63.

INDEX.

Seminal corpuscles, 62.
  vesicles, 35.
Senile impotence, 205.
Septum scroti, 32.
Sexual act, physiology of, 45.
  capacity, 50.
  excess, 137.
    atrophy of testicles from, 159.
    consequences of, 152.
    death from, 159.
    reasons for, 143.
  excitement, cause of, 48.
  function, importance of, 19.
  gymnastics, 146.
  instinct, seat of, 50.
  life, regulation of, 270.
  maturity, 45.
  organs, acquired defects of, 93.
    congenital malformations of, 87.
    disease of, and virility, 108.
  orgasm, 47.
  taste, changes in, 154.
  virility, definition of, 28.
Sinapisms, 261.
Sinus Morgagni, 40.
  prostaticus, 40.
Sitz-baths, 247.
Skopti, 93.
Sledge, 265.
Smallness of the penis, 88.
Smoking, effect on virility, 116.
Snuff-taking, effect on virility, 117.
Sodium nitrate, effect on virility, 121.
  in spermatorrhea, 234.
Sodomy, 133.
Spermatic canals, 33.
Spermatoblasts, 61.
Spermatorrhea, 174.
  treatment of, 230.
Spermatozoa, movements of, 78.
  structure of, 62.
  vitality of, 71.

Spermin, 278.
Spinal-cord disease and virility, 102.
Sponge-bath, 247.
Static electricity, 254.
Stimulants in treatment of impotence, 243.
Strictures of the urethra and virility, 111.
Suppositories, urethral, 259.
Surgical operations, 261.
Suspension, 274.
Swedish movements, 262.

Tabes dorsalis and impotence, 101.
Tea, effect on virility, 116.
Temporary impotence, 198.
Testicle, absence of one, 95.
Testicles, absence of, 94.
  anatomy of, 30.
  atrophy of, from onanism, 174.
  cancer of, and virility, 112.
  loss of, 93.
  tuberculosis of, and virility, 112.
  undescended, 92.
Tincture of veratrum viride in spermatorrhea, 234.
Tobacco, effect on virility, 116.
Travelling in treatment of impotence, 262.
Treatment, 226.
  local, 254.
  psychical, 228.
Tuberculosis of the testicles and virility, 112.
Tunica albuginea, 33.
  dartos, 31.
  vaginalis communis, 31.
  propria, 31.

Uncleanliness and onanism, 170.
Urethra, 38.
Urethral endoscope, 257.
  injections, 250.
  strictures and virility, 111.
  suppositories, 259.

Urine, incontinence of, and sexual weakness, 124.
Urnings, 132.
Utriculus prostaticus, 40.

VALERIAN, 243.
Vapor-baths, 248.
Varicocele and impotence, 113.
Vas deferens, 35.
 epididymis, 34.
Vasa aberrantia, 34.
 seminalia, 33.

Venery, excessive impotence from, 137.
Ventouse, 264.
Verumontanum, 39.
Vesicula prostatica, 40.
Virility, conditions influencing, 79.
 definition of, 28.
 importance of, 19.

WINE, effect on virility, 115.
Winternitz's rectal cooling pouch, 250.
Work and sexual power, 203.

THE END.

Medical and Surgical Works

PUBLISHED BY
W. B. SAUNDERS, 925 Walnut Street, Philadelphia, Pa.

	PAGE
Abbott on Transmissible Diseases	40
American Pocket Medical Dictionary	31
*American Text-Book of Applied Therapeutics	6
*American Text-Book of Chemistry	40
*American Text-Book of Diagnosis	40
*American Text-Book of Dis. of Children	11
*An American Text-Book of Diseases of the Eye, Ear, Nose, and Throat	13
*An American Text-Book of Genito-Urinary and Skin Diseases	12
*American Text-Book of Gynecology	10
*American Text-Book of Legal Medicine	40
*American Text-Book of Obstetrics	7
*American Text-Book of Pathology	40
*American Text-Book of Physiology	5
*American Text-Book of Practice	8
*American Text-Book of Surgery	9
Anders' Theory and Practice of Medicine	17
Ashton's Obstetrics	39
Atlas of Skin Diseases	24
Ball's Bacteriology	39
Bastin's Laboratory Exercises in Botany	32
Beck's Surgical Asepsis	37
Boisliniere's Obstetric Accidents	35
Brockway's Physics	39
Burr's Nervous Diseases	37
Butler's Materia Medica and Therapeutics	20
Cerna's Notes on the Newer Remedies	28
Chapin's Compendium of Insanity	31
Chapman's Medical Jurisprudence	37
Church and Peterson's Nervous and Mental Diseases	15
Clarkson's Histology	29
Cohen and Eshner's Diagnosis	39
Corwin's Diagnosis of the Thorax	33
Cragin's Gynæcology	39
Crookshank's Text-Book of Bacteriology	23
DaCosta's Manual of Surgery	19
De Schweinitz's Diseases of the Eye	25
Dorland's Pocket Medical Dictionary	31
Dorland's Obstetrics	37
Frothingham's Bacteriological Guide	26
Garrigues' Diseases of Women	30
Gleason's Diseases of the Ear	39
*Gould and Pyle's Curiosities of Medicine	15
Grafstrom's Massage	24
Griffith's Care of the Baby	34
Griffith's Infant's Weight Chart	35
Gross' Autobiography	22
Hampton's Nursing	35
Hare's Physiology	39
Hart's Diet in Sickness and in Health	32
Haynes' Manual of Anatomy	37
Heisler's Embryology	40
Hirst's Obstetrics	16
Hyde's Syphilis and Venereal Diseases	37
International Text-Book of Surgery	40
Jackson's Diseases of the Eye	40
Jackson and Gleason's Diseases of the Eye, Nose, and Throat	39
Keating's Pronouncing Dictionary	22
Keating's Life Insurance	35
Keen's Operation Blanks	32
Keen's Surgery of Typhoid Fever	18

	PAGE
Kyle's Diseases of Nose and Throat	40
Laine's Temperature Charts	28
Lockwood's Practice of Medicine	37
Long's Syllabus of Gynecology	30
Macdonald's Surgical Diagnosis and Treatment	18
McFarland's Pathogenic Bacteria	26
Mallory and Wright's Pathological Technique	18
Martin's Surgery	39
Martin's Minor Surgery, Bandaging, and Venereal Diseases	39
Meigs' Feeding in Early Infancy	26
Moore's Orthopedic Surgery	19
Morris' Materia Medica and Therapeutics	39
Morris' Practice of Medicine	39
Morten's Nurses' Dictionary	34
Nancrede's Anatomy and Dissection	27
Nancrede's Anatomy	39
Norris' Syllabus of Obstetrical Lectures	33
Penrose's Diseases of Women	20
Powell's Diseases of Children	39
Pryor's Pelvic Inflammations	40
Pye's Bandaging and Surgical Dressing	19
Raymond's Physiology	37
Rowland's Clinical Skiagraphy	29
Saundby's Renal and Urinary Diseases	21
*Saunders' American Year-Book of Medicine and Surgery	14
Saunders' Medical Hand-Atlases	3, 4
Saunders' Pocket Medical Formulary	31
Saunders' New Series of Manuals	36, 37
Saunders' Series of Question Compends	38, 39
Sayre's Practice of Pharmacy	39
Semple's Pathology and Morbid Anatomy	39
Semple's Legal Medicine, Toxicology, and Hygiene	39
Senn's Genito-Urinary Tuberculosis	20
Senn's Tumors	21
Senn's Syllabus of Lectures on Surgery	33
Shaw's Nervous Diseases and Insanity	39
Starr's Diet-Lists for Children	34
Stelwagon's Diseases of the Skin	39
Stengel's Pathology	16
Stevens' Materia Medica and Therapeutics	28
Stevens' Practice of Medicine	27
Stewart's Manual of Physiology	33
Stewart and Lawrance's Medical Electricity	39
Stoney's Materia Medica for Nurses	27
Stoney's Practical Points in Nursing	23
Sutton and Giles' Diseases of Women	25, 37
Thomas's Diet-List and Sick-Room Dietary	34
Thornton's Dose-Book and Manual of Prescription-Writing	37
Thresh's Water and Water Supplies	29
Van Valzah and Nisbet's Diseases of the Stomach	17
Vecki's Sexual Impotence	29
Vierordt and Stuart's Medical Diagnosis	24
Warren's Surgical Pathology	21
Wolff's Chemistry	39
Wolff's Examination of Urine	30

GENERAL INFORMATION.

One Price. One price absolutely without deviation. No discounts allowed, regardless of the number of books purchased at one time. Prices on all works have been fixed extremely low, with the view to selling them strictly net and for cash.

Orders. An order accompanied by remittance will receive prompt attention, books being sent to any address in the United States, by mail or express, all charges prepaid. We prefer to send books **by express** when possible.

Cash or Credit. To physicians of approved credit who furnish satisfactory references our books will be sent free of C. O. D. **One volume or two** on thirty days' time if credit is desired; larger purchases on monthly payment plan. See offer below.

How to Send Money by Mail. There are four ways by which money can be sent at our risk, namely: a post-office money order, an express money order, a bank-check (draft), and in a registered letter. Money sent in any other way is at the sender's risk. Silver should not be sent through the mail.

Shipments. All books, being packed in patent metal-edged boxes, necessarily reach our patrons by mail or express in excellent condition.

Subscription Books. Books in this catalogue marked with a star (*) are for sale by subscription only, and may be secured by ordering them through any of our authorized travelling salesmen, or direct from the Philadelphia office; they are **not** for sale by booksellers. All other books in our catalogue can be procured of any bookseller at the advertised price, or directly from us.

Miscellaneous Books. We carry in stock only our own publications, but can supply the publications of other houses (except subscription books) on receipt of publisher's price.

Latest Editions. In every instance the latest revised edition is sent.

Bindings. In ordering, be careful to state the style of binding desired— Cloth, Sheep, or Half Morocco.

Special Offer. Monthly Payment Plan. To physicians of approved credit who furnish satisfactory references books will be sent express prepaid; terms, $5.00 cash upon delivery of books, and monthly payments of $5.00 thereafter until full amount is paid. Any of the publications of W. B. Saunders (100 titles to select from) may be had in this way at catalogue price, including the American Text-Book Series, the Medical Hand-Atlases, etc. All payments to be made by mail or otherwise, free of all expense to us.

SAUNDERS'
MEDICAL HAND-ATLASES.

THE series of books included under this title consists of authorized translations into English of the world-famous **Lehmann Medicinische Handatlanten,** which for **scientific accuracy, pictorial beauty, compactness, and cheapness** surpass any similar volumes ever published.

Each volume contains from **50 to 100 colored plates,** executed by the most skilful German lithographers, besides numerous illustrations in the text. There is a full and appropriate **description,** and each book contains a condensed but adequate **outline of the subject** to which it is devoted.

One of the most valuable features of these atlases is that they offer a **ready and satisfactory substitute for clinical observation.** Such observation, of course, is available only to the residents in large medical centers; and even then the requisite variety is seen only after long years of routine hospital work. To those unable to attend important clinics these books will be absolutely indispensable, as presenting in a complete and convenient form the most accurate reproductions of clinical work, interpreted by the most competent of clinical teachers.

While appreciating the value of such colored plates, the profession has heretofore been practically debarred from purchasing similar works because of their extremely high price, made necessary by a limited sale and an enormous expense of production. In planning this series, however, arrangements were made with representative publishers in the chief medical centers of the world for the publication of translations of the atlases into nine different languages, the lithographic plates for all being made in Germany, where work of this kind has been brought to the greatest perfection. The enormous expense of making the plates being shared by the various publishers, the cost to each one was reduced to practically one-tenth. Thus by reason of their **universal translation** and reproduction, affording international distribution, the publishers have been enabled to secure for these atlases the **best artistic and professional talent,** to produce them in the **most elegant style,** and yet to offer them at a **price heretofore unapproached in cheapness.** The great success of the undertaking is demonstrated by the fact that the volumes have already appeared in **nine different languages**—German, English, French, Italian, Russian, Spanish, Danish, Swedish, and Hungarian.

In view of the unprecedented success of these works, Mr. Saunders has contracted with the publisher of the original German edition for **one hundred thousand copies** of the atlases. In consideration of this enormous undertaking, the publisher has been enabled to prepare and furnish special additional colored plates, making the series even **handsomer and more complete** than was originally intended.

As an indication of the great practical value of the atlases and of the immense favor with which they have been received, it should be noted that the **Medical Department of the U. S. Army** has adopted the "**Atlas of Operative Surgery**" as its standard, and has ordered the book in large quantities for distribution to the various regiments and army posts.

The same careful and competent **editorial supervision** has been secured in the English edition as in the originals. The translations have been edited by the **leading American specialists** in the different subjects. The volumes are of a uniform and convenient size ($5 \times 7\frac{1}{2}$ inches), and are substantially bound.

(*For List of Volumes in this Series, see next page.*)

SAUNDERS' MEDICAL HAND-ATLASES.

VOLUMES NOW READY.

Atlas of Internal Medicine and Clinical Diagnosis. By DR. CHR. JAKOB, of Erlangen. Edited by AUGUSTUS A. ESHNER, M. D., Professor of Clinical Medicine Philadelphia Polyclinic. 68 colored plates, 64 text illustrations, and 259 pages of text. Cloth, $3.00 net.

"The charm of the book is its clearness, conciseness, and the accuracy and beauty of its illustrations. It deals with facts. It vividly illustrates those facts. It is a scientific work put together for ready reference."—*Brooklyn Medical Journal.*

Atlas of Legal Medicine. By DR. E. R. VON HOFMANN, of Vienna. Edited by FREDERICK PETERSON, M. D., Clinical Professor of Mental Diseases, Woman's Medical College, New York; Chief of Clinic, Nervous Dept., College of Physicians and Surgeons, New York. With 120 colored figures on 56 plates, and 193 beautiful half-tone illustrations. Cloth, $3.50 net.

"Hofmann's 'Atlas of Legal Medicine' is a unique work. This immense field finds in this book a pictorial presentation that far excels anything with which we are familiar in any other work."—*Philadelphia Medical Journal.*

Atlas of Diseases of the Larynx. By DR. L. GRÜNWALD, of Munich. Edited by CHARLES P. GRAYSON, M. D., Physician-in-Charge, Throat and Nose Department, Hospital of the University of Pennsylvania. With 107 colored figures on 44 plates, 25 text-illustrations, and 103 pages of text. Cloth, $2.50 net.

"Aided as it is by magnificently executed illustrations in color, it cannot fail of being of the greatest advantage to students, general practitioners, and expert laryngologists."—*St. Louis Medical and Surgical Journal.*

Atlas of Operative Surgery. By DR. O. ZUCKERKANDL, of Vienna. Edited by J. CHALMERS DACOSTA, M. D., Clinical Professor of Surgery, Jefferson Medical College, Philadelphia. With 24 colored plates, 217 text-illustrations, and 395 pages of text. Cloth, $3.00 net.

"We know of no other work that combines such a wealth of beautiful illustrations with clearness and conciseness of language, that is so entirely abreast of the latest achievements, and so useful both for the beginner and for one who wishes to increase his knowledge of operative surgery."—*Münchener medicinische Wochenschrift.*

Atlas of Syphilis and the Venereal Diseases. By PROF. DR. FRANZ MRACEK, of Vienna. Edited by L. BOLTON BANGS, M. D., Professor of Genito-Urinary Surgery, University and Bellevue Hospital Medical College, New York. With 71 colored plates from original water-colors, 16 black-and-white illustrations, and 122 pages of text. Cloth, $3.50 net.

"A glance through the book is almost like actual attendance upon a famous clinic."—*Journal of the American Medical Association.*

Atlas of External Diseases of the Eye. By DR. O. HAAB, of Zurich. Edited by G. E. DE SCHWEINITZ, M. D., Professor of Ophthalmology, Jefferson Medical College, Philadelphia With 76 colored illustrations on 40 plates, and 228 pages of text. Cloth, $3.00 net.

Atlas of Skin Diseases. By PROF. DR. FRANZ MRACEK, of Vienna. Edited by HENRY W. STELWAGON, M. D., Clinical Professor of Dermatology, Jefferson Medical College, Philadelphia. With 63 colored plates, 39 beautiful half-tone illustrations, and 200 pages of text. Cloth, $3.50 net.

IN PREPARATION.

Atlas of Pathological Histology.
Atlas of Orthopedic Surgery.
Atlas of General Surgery.

Atlas of Operative Gynecology.
Atlas of Psychiatry.
Atlas of Diseases of the Ear.

*AN AMERICAN TEXT-BOOK OF PHYSIOLOGY.

Edited by WILLIAM H. HOWELL, PH. D., M. D., Professor of Physiology in the Johns Hopkins University, Baltimore, Md. One handsome octavo volume of 1052 pages, fully illustrated. Prices: Cloth, $6.00 net; Sheep or Half-Morocco, $7.00 net.

This work is the most notable attempt yet made in America to combine in one volume the entire subject of Human Physiology by well-known teachers who have given especial study to that part of the subject upon which they write. The completed work represents the present status of the science of Physiology, particularly from the standpoint of the student of medicine and of the medical practitioner.

The collaboration of several teachers in the preparation of an elementary text-book of physiology is unusual, the almost invariable rule heretofore having been for a single author to write the entire book. One of the advantages to be derived from this collaboration method is that the more limited literature necessary for consultation by each author has enabled him to base his elementary account upon a comprehensive knowledge of the subject assigned to him; another, and perhaps the most important, advantage is that the student gains the point of view of a number of teachers. In a measure he reaps the same benefit as would be obtained by following courses of instruction under different teachers. The different standpoints assumed, and the differences in emphasis laid upon the various lines of procedure, chemical, physical, and anatomical, should give the student a better insight into the methods of the science as it exists to-day. The work will also be found useful to many medical practitioners who may wish to keep in touch with the development of modern physiology.

CONTRIBUTORS:

HENRY P. BOWDITCH, M. D.,
Professor of Physiology, Harvard Medical School.

JOHN G. CURTIS, M. D.,
Professor of Physiology, Columbia University, N. Y. (College of Physicians and Surgeons).

HENRY H. DONALDSON, Ph. D.,
Head-Professor of Neurology, University of Chicago.

W. H. HOWELL, Ph. D., M. D.,
Professor of Physiology, Johns Hopkins University.

FREDERIC S. LEE, Ph. D.,
Adjunct Professor of Physiology, Columbia University, N. Y. (College of Physicians and Surgeons).

WARREN P. LOMBARD, M. D.,
Professor of Physiology, University of Michigan.

GRAHAM LUSK, Ph. D.,
Professor of Physiology, Yale Medical School.

W. T. PORTER, M. D.,
Assistant Professor of Physiology, Harvard Medical School.

EDWARD T. REICHERT, M. D.,
Professor of Physiology, University of Pennsylvania.

HENRY SEWALL, Ph. D., M. D.,
Professor of Physiology, Medical Department, University of Denver.

" We can commend it most heartily, not only to all students of physiology, but to every physician and pathologist, as a valuable and comprehensive work of reference, written by men who are of eminent authority in their own special subjects."—*London Lancet.*

" To the practitioner of medicine and to the advanced student this volume constitutes, we believe, the best exposition of the present status of the science of physiology in the English language."—*American Journal of the Medical Sciences.*

*AN AMERICAN TEXT-BOOK OF APPLIED THERAPEU-
TICS. For the Use of Practitioners and Students. Edited by
JAMES C. WILSON, M. D., Professor of the Practice of Medicine and of
Clinical Medicine in the Jefferson Medical College. One handsome octavo
volume of 1326 pages. Illustrated. Prices: Cloth, $7.00 net; Sheep or
Half-Morocco, $8.00 net.

The arrangement of this volume has been based, so far as possible, upon modern pathologic doctrines, beginning with the intoxications, and following with infections, diseases due to internal parasites, diseases of undetermined origin, and finally the disorders of the several bodily systems—digestive, respiratory, circulatory, renal, nervous, and cutaneous. It was thought proper to include also a consideration of the disorders of pregnancy.

The articles, with two exceptions, are the contributions of American writers. Written from the standpoint of the practitioner, the aim of the work is to facilitate the application of knowledge to the prevention, the cure, and the alleviation of disease. The endeavor throughout has been to conform to the title of the book—Applied Therapeutics—to indicate the course of treatment to be pursued at the bedside, rather than to name a list of drugs that have been used at one time or another.

The list of contributors comprises the names of many who have acquired distinction as practitioners and teachers of practice, of clinical medicine, and of the specialties.

CONTRIBUTORS:

Dr. I. E. Atkinson, Baltimore, Md.
Sanger Brown, Chicago, Ill.
John B. Chapin, Philadelphia, Pa.
William C. Dabney, Charlottesville, Va.
John Chalmers DaCosta, Philada., Pa.
I. N. Danforth, Chicago, Ill.
John L. Dawson, Jr., Charleston, S. C.
F. X. Dercum, Philadelphia, Pa.
George Dock, Ann Arbor, Mich.
Robert T. Edes, Jamaica Plain, Mass.
Augustus A. Eshner, Philadelphia, Pa.
J. T. Eskridge, Denver, Col.
F. Forchheimer, Cincinnati, O.
Carl Frese, Philadelphia, Pa.
Edwin E. Graham, Philadelphia, Pa.
John Guitéras, Philadelphia, Pa.
Frederick P. Henry, Philadelphia, Pa.
Guy Hinsdale, Philadelphia, Pa.
Orville Horwitz, Philadelphia, Pa.
W. W. Johnston, Washington, D. C.
Ernest Laplace, Philadelphia, Pa.
A. Laveran, Paris, France.

Dr. James Hendrie Lloyd, Philadelphia, Pa.
John Noland Mackenzie, Baltimore, Md.
J. W. McLaughlin, Austin, Texas.
A. Lawrence Mason, Boston, Mass.
Charles K. Mills, Philadelphia, Pa.
John K. Mitchell, Philadelphia, Pa.
W. P. Northrup, New York City.
William Osler, Baltimore, Md.
Frederick A. Packard, Philadelphia, Pa.
Theophilus Parvin, Philadelphia, Pa.
Beaven Rake, London, England.
E. O. Shakespeare, Philadelphia, Pa.
Wharton Sinkler, Philadelphia, Pa.
Louis Starr, Philadelphia, Pa.
Henry W. Stelwagon, Philadelphia, Pa.
James Stewart, Montreal, Canada.
Charles G. Stockton, Buffalo, N. Y.
James Tyson, Philadelphia, Pa.
Victor C. Vaughan, Ann Arbor, Mich.
James T. Whittaker, Cincinnati, O.
J. C. Wilson, Philadelphia, Pa.

" As a work either for study or reference it will be of great value to the practitioner, as it is virtually an exposition of such clinical therapeutics as experience has taught to be of the most value. Taking it all in all, no recent publication on therapeutics can be compared with this one in practical value to the working physician."—*Chicago Clinical Review.*

"The whole field of medicine has been well covered. The work is thoroughly practical, and while it is intended for practitioners and students, it is a better book for the general practitioner than for the student. The young practitioner especially will find it extremely suggestive and helpful."—*The Indian Lancet.*

*AN AMERICAN TEXT-BOOK OF OBSTETRICS. Edited by RICHARD C. NORRIS, M. D.; Art Editor, ROBERT L. DICKINSON, M. D. One handsome octavo volume of over 1000 pages, with nearly 900 colored and half-tone illustrations. Prices: Cloth, $7.00; Sheep or Half-Morocco, $8.00.

The advent of each successive volume of the *series* of the AMERICAN TEXT-BOOKS has been signalized by the most flattering comment from both the Press and the Profession. The high consideration received by these text-books, and their attainment to an authoritative position in current medical literature, have been matters of deep *international* interest, which finds its fullest expression in the demand for these publications from all parts of the civilized world.

In the preparation of the "AMERICAN TEXT-BOOK OF OBSTETRICS" the editor has called to his aid proficient collaborators whose professional prominence entitles them to recognition, and whose disquisitions exemplify **Practical Obstetrics**. While these writers were each assigned special themes for discussion, the correlation of the subject-matter is, nevertheless, such as ensures logical connection in treatment, the deductions of which thoroughly represent the latest advances in the science, and which elucidate *the best modern methods of procedure*.

The more conspicuous feature of the treatise is its wealth of illustrative matter. The production of the illustrations had been in progress for several years, under the personal supervision of Robert L. Dickinson, M. D., to whose artistic judgment and professional experience is due the **most sumptuously illustrated work of the period**. By means of the photographic art, combined with the skill of the artist and draughtsman, conventional illustration is superseded by rational methods of delineation.

Furthermore, the volume is a revelation as to the possibilities that may be reached in mechanical execution, through the unsparing hand of its publisher.

CONTRIBUTORS:

Dr. James C. Cameron.
Edward P. Davis.
Robert L. Dickinson.
Charles Warrington Earle.
James H. Etheridge.
Henry J. Garrigues.
Barton Cooke Hirst.
Charles Jewett.

Dr. Howard A. Kelly.
Richard C. Norris.
Chauncey D. Palmer.
Theophilus Parvin.
George A. Piersol.
Edward Reynolds.
Henry Schwarz.

"At first glance we are overwhelmed by the magnitude of this work in several respects, viz.: First, by the size of the volume, then by the array of eminent teachers in this department who have taken part in its production, then by the profuseness and character of the illustrations, and last, but not least, the conciseness and clearness with which the text is rendered. This is an entirely new composition, embodying the highest knowledge of the art as it stands to-day by authors who occupy the front rank in their specialty, and there are many of them. We cannot turn over these pages without being struck by the superb illustrations which adorn so many of them. We are confident that this most practical work will find instant appreciation by practitioners as well as students."—*New York Medical Times*.

Permit me to say that your American Text-Book of Obstetrics is the most magnificent medical work that I have ever seen. I congratulate you and thank you for this superb work, which alone is sufficient to place you first in the ranks of medical publishers.
With profound respect I am sincerely yours, ALEX. J. C. SKENE.

*AN AMERICAN TEXT-BOOK OF THE THEORY AND PRACTICE OF MEDICINE. By American Teachers. Edited by WILLIAM PEPPER, M. D., LL.D., Provost and Professor of the Theory and Practice of Medicine and of Clinical Medicine in the University of Pennsylvania. Complete in two handsome royal-octavo volumes of about 1000 pages each, with illustrations to elucidate the text wherever necessary. Price per Volume: Cloth, $5.00 net; Sheep or Half-Morocco, $6.00 net.

VOLUME I. CONTAINS:

Hygiene.—Fevers (Ephemeral, Simple Continued, Typhus, Typhoid, Epidemic Cerebrospinal Meningitis, and Relapsing).—Scarlatina, Measles, Rötheln, Variola, Varioloid, Vaccinia, Varicella, Mumps, Whooping-cough, Anthrax, Hydrophobia, Trichinosis, Actinomycosis, Glanders, and Tetanus.—Tuberculosis, Scrofula, Syphilis, Diphtheria, Erysipelas, Malaria, Cholera, and Yellow Fever.—Nervous, Muscular, and Mental Diseases etc.

VOLUME II. CONTAINS:

Urine (Chemistry and Microscopy).—Kidney and Lungs.—Air-passages (Larynx and Bronchi) and Pleura.—Pharynx, Œsophagus, Stomach and Intestines (including Intestinal Parasites), Heart, Aorta, Arteries and Veins.—Peritoneum, Liver, and Pancreas.—Diathetic Diseases (Rheumatism, Rheumatoid Arthritis, Gout, Lithæmia, and Diabetes).—Blood and Spleen.—Inflammation, Embolism, Thrombosis, Fever, and Bacteriology.

The articles are not written as though addressed to students in lectures, but are exhaustive descriptions of diseases, with the newest facts as regards Causation, Symptomatology, Diagnosis, Prognosis, and Treatment, including a large number of approved formulæ. The recent advances made in the study of the bacterial origin of various diseases are fully described, as well as the bearing of the knowledge so gained upon prevention and cure. The subjects of Bacteriology as a whole and of Immunity are fully considered in a separate section.

Methods of diagnosis are given the most minute and careful attention, thus enabling the reader to learn the very latest methods of investigation without consulting works specially devoted to the subject.

CONTRIBUTORS:

Dr. J. S. Billings, Philadelphia.
Francis Delafield, New York.
Reginald H. Fitz, Boston.
James W. Holland, Philadelphia.
Henry M. Lyman, Chicago.
William Osler, Baltimore.

Dr. William Pepper, Philadelphia.
W. Gilman Thompson, New York.
W. H. Welch, Baltimore.
James T. Whittaker, Cincinnati.
James C. Wilson, Philadelphia.
Horatio C. Wood, Philadelphia.

"We reviewed the first volume of this work, and said: 'It is undoubtedly one of the best text-books on the practice of medicine which we possess.' A consideration of the second and last volume leads us to modify that verdict and to say that the completed work is, in our opinion, THE BEST of its kind it has ever been our fortune to see. It is complete, thorough, accurate, and clear. It is well written, well arranged, well printed, well illustrated, and well bound. It is a model of what the modern text-book should be."—*New York Medical Journal.*

"A library upon modern medical art. The work must promote the wider diffusion of sound knowledge."—*American Lancet.*

"A trusty counsellor for the practitioner or senior student, on which he may implicitly rely."—*Edinburgh Medical Journal.*

*AN AMERICAN TEXT-BOOK OF SURGERY. Edited by WILLIAM W. KEEN, M. D., LL.D., and J. WILLIAM WHITE, M. D., PH. D. Forming one handsome royal octavo volume of 1230 pages (10 × 7 inches), with 496 wood-cuts in text, and 37 colored and half-tone plates, many of them engraved from original photographs and drawings furnished by the authors. Price: Cloth, $7.00 net; Sheep or Half Morocco, $8.00 net.

THIRD EDITION, THOROUGHLY REVISED.

The want of a text-book which could be used by the practitioner and at the same time be recommended to the medical student has been deeply felt, especially by teachers of surgery; hence, when it was suggested to a number of these that it would be well to unite in preparing a text-book of this description, great unanimity of opinion was found to exist, and the gentlemen below named gladly consented to join in its production. While there is no distinctive American Surgery, yet America has contributed very largely to the progress of modern surgery, and among the foremost of those who have aided in developing this art and science will be found the authors of the present volume. All of them are teachers of surgery in leading medical schools and hospitals in the United States and Canada.

Especial prominence has been given to Surgical Bacteriology, a feature which is believed to be unique in a surgical text-book in the English language. Asepsis and Antisepsis have received particular attention. The text is brought well up to date in such important branches as cerebral, spinal, intestinal, and pelvic surgery, the most important and newest operations in these departments being described and illustrated.

The text of the entire book has been submitted to all the authors for their mutual criticism and revision—an idea in book-making that is entirely new and original. The book as a whole, therefore, expresses on all the important surgical topics of the day the consensus of opinion of the eminent surgeons who have joined in its preparation.

One of the most attractive features of the book is its illustrations. Very many of them are original and faithful reproductions of photographs taken directly from patients or from specimens.

CONTRIBUTORS:

Dr. Phineas S. Conner, Cincinnati.
Frederic S. Dennis, New York.
William W. Keen, Philadelphia.
Charles B. Nancrede, Ann Arbor, Mich.
Roswell Park, Buffalo, New York.
Lewis S. Pilcher, New York.

Dr. Nicholas Senn, Chicago.
Francis J. Shepherd, Montreal, Canada.
Lewis A. Stimson, New York.
J. Collins Warren, Boston.
J. William White, Philadelphia.

"If this text-book is a fair reflex of the present position of American surgery, we must admit it is of a very high order of merit, and that English surgeons will have to look very carefully to their laurels if they are to preserve a position in the van of surgical practice."—*London Lancet.*

*AN AMERICAN TEXT-BOOK OF GYNECOLOGY, MEDICAL AND SURGICAL, for the use of Students and Practitioners. Edited by J. M. BALDY, M. D. Forming a handsome royal-octavo volume of 718 pages, with 341 illustrations in the text and 38 colored and half-tone plates. Prices: Cloth, $6.00 net; Sheep or Half-Morocco, $7.00 net.

SECOND EDITION, THOROUGHLY REVISED.

In this volume all anatomical descriptions, excepting those essential to a clear understanding of the text, have been omitted, the illustrations being largely depended upon to elucidate the anatomy of the parts. This work, which is thoroughly practical in its teachings, is intended, as its title implies, to be a working text-book for physicians and students. A clear line of treatment has been laid down in every case, and although no attempt has been made to discuss mooted points, still the most important of these have been noted and explained. The operations recommended are fully illustrated, so that the reader, having a picture of the procedure described in the text under his eye, cannot fail to grasp the idea. All extraneous matter and discussions have been carefully excluded, the attempt being made to allow no unnecessary details to cumber the text. The subject-matter is brought up to date at every point, and the work is as nearly as possible the combined opinions of the ten specialists who figure as the authors.

In the revised edition much new material has been added, and some of the old eliminated or modified. More than forty of the old illustrations have been replaced by new ones, which add very materially to the elucidation of the text, as they picture methods, not specimens. The chapters on technique and after-treatment have been considerably enlarged, and the portions devoted to plastic work have been so greatly improved as to be practically new. Hysterectomy has been rewritten, and all the descriptions of operative procedures have been carefully revised and fully illustrated.

CONTRIBUTORS:

Dr. Henry T. Byford.
John M. Baldy.
Edwin Cragin.
J. H. Etheridge.
William Goodell.

Dr. Howard A. Kelly.
Florian Krug.
E. E. Montgomery.
William R. Pryor.
George M. Tuttle.

"The most notable contribution to gynecological literature since 1887, ... and the most complete exponent of gynecology which we have. No subject seems to have been neglected, ... and the gynecologist and surgeon, and the general practitioner who has any desire to practise diseases of women, will find it of practical value. In the matter of illustrations and plates the book surpasses anything we have seen."—*Boston Medical and Surgical Journal.*

"A thoroughly modern text-book, and gives reliable and well-tempered advice and instruction."—*Edinburgh Medical Journal.*

"The harmony of its conclusions and the homogeneity of its style give it an individuality which suggests a single rather than a multiple authorship."—*Annals of Surgery.*

"It must command attention and respect as a worthy representation of our advanced clinical teaching."—*American Journal of Medical Sciences.*

*AN AMERICAN TEXT-BOOK OF THE DISEASES OF CHILDREN. By American Teachers. Edited by LOUIS STARR, M. D., assisted by THOMPSON S. WESTCOTT, M. D. In one handsome royal-8vo volume of 1244 pages, profusely illustrated with wood-cuts, half-tone and colored plates. Net Prices: Cloth, $7.00; Sheep or Half-Morocco, $8.00.

SECOND EDITION, REVISED AND ENLARGED.

The plan of this work embraces a series of original articles written by some sixty well-known pædiatrists, representing collectively the teachings of the most prominent medical schools and colleges of America. The work is intended to be a PRACTICAL book, suitable for constant and handy reference by the practitioner and the advanced student.

Especial attention has been given to the latest accepted teachings upon the etiology, symptoms, pathology, diagnosis, and treatment of the disorders of children, with the introduction of many special formulæ and therapeutic procedures.

In this new edition the whole subject matter has been carefully revised, new articles added, some original papers emended, and a number entirely rewritten. The new articles include "Modified Milk and Percentage Milk-Mixtures," "Lithemia," and a section on "Orthopedics." Those rewritten are "Typhoid Fever," "Rubella," "Chicken-pox," "Tuberculous Meningitis," "Hydrocephalus," and "Scurvy;" while extensive revision has been made in "Infant Feeding," "Measles," "Diphtheria," and "Cretinism." The volume has thus been much increased in size by the introduction of fresh material.

CONTRIBUTORS:

Dr. S. S. Adams, Washington.
John Ashhurst, Jr., Philadelphia.
A. D. Blackader, Montreal, Canada.
David Bovaird, New York.
Dillon Brown, New York.
Edward M. Buckingham, Boston.
Charles W. Burr, Philadelphia.
W. E. Casselberry, Chicago.
Henry Dwight Chapin, New York.
W. S. Christopher, Chicago.
Archibald Church, Chicago.
Floyd M. Crandall, New York.
Andrew F. Currier, New York.
Roland G. Curtin, Philadelphia
J. M. DaCosta, Philadelphia.
I. N. Danforth, Chicago.
Edward P. Davis, Philadelphia.
John B. Deaver, Philadelphia.
G. E. de Schweinitz, Philadelphia.
John Dorning, New York.
Charles Warrington Earle, Chicago.
Wm. A. Edwards, San Diego, Cal.
F. Forchheimer, Cincinnati.
J. Henry Fruitnight, New York.
J. P. Crozer Griffith, Philadelphia.
W. A. Hardaway. St. Louis.
M. P Hatfield, Chicago.
Barton Cooke Hirst, Philadelphia.
H. Illoway, Cincinnati.
Henry Jackson, Boston.
Charles G. Jennings, Detroit.
Henry Koplik, New York.

Dr. Thomas S. Latimer, Baltimore.
Albert R. Leeds, Hoboken, N. J.
J. Hendrie Lloyd, Philadelphia.
George Roe Lockwood, New York.
Henry M. Lyman, Chicago.
Francis T. Miles, Baltimore.
Charles K Mills, Philadelphia.
James E. Moore, Minneapolis.
F. Gordon Morrill, Boston.
John H. Musser, Philadelphia.
Thomas R. Neilson, Philadelphia.
W. P. Northrup, New York.
William Osler, Baltimore.
Frederick A. Packard, Philadelphia.
William Pepper, Philadelphia.
Frederick Peterson, New York.
W. T. Plant, Syracuse, New York.
William M. Powell, Atlantic City.
B. K. Rachford, Cincinnati.
B. Alexander Randall, Philadelphia.
Edward O. Shakespeare, Philadelphia
F. C. Shattuck, Boston.
J. Lewis Smith, New York.
Louis Starr, Philadelphia.
M. Allen Starr, New York.
Charles W. Townsend, Boston.
James Tyson, Philadelphia.
W. S. Thayer, Baltimore.
Victor C. Vaughan, Ann Arbor, Mich
Thompson S. Westcott, Philadelphia.
Henry R. Wharton, Philadelphia.
J William White, Philadelphia.
I C. Wilson, Philadelphia.

* **AN AMERICAN TEXT-BOOK OF GENITO-URINARY AND SKIN DISEASES.** By 47 Eminent Specialists and Teachers. Edited by L. BOLTON BANGS, M. D., Professor of Genito-Urinary Surgery, University and Bellevue Hospital Medical College, New York; and W. A. HARDAWAY, M. D., Professor of Diseases of the Skin, Missouri Medical College. Imperial octavo volume of 1229 pages, with 300 engravings and 20 full-page colored plates. Cloth, $7.00 net; Sheep or Half Morocco, $8.00 net.

This addition to the series of "American Text-Books," it is confidently believed, will meet the requirements of both students and practitioners, giving, as it does, a comprehensive and detailed presentation of the Diseases of the Genito-Urinary Organs, of the Venereal Diseases, and of the Affections of the Skin.

Having secured the collaboration of well-known authorities in the branches represented in the undertaking, the editors have not restricted the contributors in regard to the particular views set forth, but have offered every facility for the free expression of their individual opinions. The work will therefore be found to be original, yet homogeneous and fully representative of the several departments of medical science with which it is concerned.

CONTRIBUTORS:

Dr. Chas. W. Allen, New York.
I. E. Atkinson, Baltimore.
L. Bolton Bangs, New York.
P. R. Bolton, New York.
Lewis C. Bosher, Richmond, Va.
John T. Bowen, Boston.
J. Abbott Cantrell, Philadelphia.
William T. Corlett, Cleveland, Ohio.
B. Farquhar Curtis, New York.
Condict W. Cutler, New York.
Isadore Dyer, New Orleans.
Christian Fenger, Chicago.
John A. Fordyce, New York.
Eugene Fuller, New York.
R. H. Greene, New York.
Joseph Grindon, St. Louis.
Graeme M. Hammond, New York.
W. A. Hardaway, St. Louis.
M. B. Hartzell, Philadelphia.
Louis Heitzmann, New York.
James S. Howe, Boston.
George T. Jackson, New York.
Abraham Jacobi, New York.
James C. Johnston, New York.

Dr. Hermann G. Klotz, New York.
J. H. Linsley, Burlington, Vt.
G. F. Lydston, Chicago.
Hartwell N. Lyon, St. Louis.
Edward Martin, Philadelphia.
D. G. Montgomery, San Francisco.
James Pedersen, New York.
S. Pollitzer, New York.
Thomas R. Pooley, New York.
A. R. Robinson, New York.
A. E. Regensburger, San Francisco.
Francis J. Shepherd, Montreal, Can.
S. C. Stanton, Chicago, Ill.
Emmanuel J. Stout, Philadelphia.
Alonzo E. Taylor, Philadelphia.
Robert W. Taylor, New York.
Paul Thorndike, Boston.
H. Tuholske, St. Louis.
Arthur Van Harlingen, Philadelphia.
Francis S. Watson, Boston.
J. William White, Philadelphia.
J. McF. Winfield, Brooklyn.
Alfred C. Wood, Philadelphia.

"This voluminous work is thoroughly up to date, and the chapters on genito-urinary diseases are especially valuable. The illustrations are fine and are mostly original. The section on dermatology is concise and in every way admirable."—*Journal of the American Medical Association.*

"This volume is one of the best yet issued of the publisher's series of 'American Text-Books.' The list of contributors represents an extraordinary array of talent and extended experience. The book will easily take the place in comprehensiveness and value of the half dozen or more costly works on these subjects which have hitherto been necessary to a well-equipped library."—*New York Polyclinic.*

* **AN AMERICAN TEXT-BOOK OF DISEASES OF THE EYE, EAR, NOSE, AND THROAT.** Edited by GEORGE E. DE SCHWEINITZ, A. M., M. D., Professor of Ophthalmology, Jefferson Medical College; and B. ALEXANDER RANDALL, A. M., M. D., Clinical Professor of Diseases of the Ear, University of Pennsylvania. One handsome imperial octavo volume of 1251 pages; 766 illustrations, 59 of them colored. Prices: Cloth, $7.00 net; Sheep or Half-Morocco, $8.00 net.

Just Issued.

The present work is the only book ever published embracing diseases of the intimately related organs of the eye, ear, nose, and throat. Its special claim to favor is based on encyclopedic, authoritative, and practical treatment of the subjects.

Each section of the book has been entrusted to an author who is specially identified with the subject on which he writes, and who therefore presents his case in the manner of an expert. Uniformity is secured and overlapping prevented by careful editing and by a system of cross-references which forms a special feature of the volume, enabling the reader to come into touch with all that is said on any subject in different portions of the book.

Particular emphasis is laid on the most approved methods of treatment, so that the book shall be one to which the student and practitioner can refer for information in practical work. Anatomical and physiological problems, also, are fully discussed for the benefit of those who desire to investigate the more abstruse problems of the subject.

CONTRIBUTORS:

Dr. Henry A. Alderton, Brooklyn.
Harrison Allen, Philadelphia.
Frank Allport, Chicago.
Morris J. Asch, New York.
S. C. Ayres, Cincinnati.
R. O. Beard, Minneapolis.
Clarence J. Blake, Boston.
Arthur A. Bliss, Philadelphia.
Albert P. Brubaker, Philadelphia.
J. H. Bryan, Washington, D. C.
Albert H. Buck, New York.
F. Buller, Montreal, Can.
Swan M. Burnett, Washington, D. C.
Flemming Carrow, Ann Arbor, Mich.
W. E. Casselberry, Chicago.
Colman W. Cutler, New York.
Edward B. Dench, New York.
William S. Dennett, New York.
George E. de Schweinitz, Philadelphia.
Alexander Duane, New York.
John W. Farlow, Boston, Mass.
Walter J. Freeman, Philadelphia.
H. Gifford, Omaha, Neb.
W. C. Glasgow, St. Louis.
J. Orne Green, Boston.
Ward A. Holden, New York.
Christian R. Holmes, Cincinnati.
William E. Hopkins, San Francisco.
F. C. Hotz, Chicago.
Lucien Howe, Buffalo, N. Y.

Dr. Alvin A. Hubbell, Buffalo, N. Y.
Edward Jackson, Philadelphia.
J. Ellis Jennings, St. Louis.
Herman Knapp, New York.
Chas. W. Kollock, Charleston, S. C.
G. A. Leland, Boston.
J. A. Lippincott, Pittsburg, Pa.
G. Hudson Makuen, Philadelphia.
John H. McCollom, Boston.
H. G. Miller, Providence, R. I.
B. L. Milliken, Cleveland, Ohio.
Robert C. Myles, New York.
James E. Newcomb, New York.
R. J. Phillips, Philadelphia.
George A. Piersol, Philadelphia.
W. P. Porcher, Charleston, S. C.
B. Alex. Randall, Philadelphia.
Robert L. Randolph, Baltimore.
John O. Roe, Rochester, N. Y.
Charles E. de M. Sajous, Philadelphia.
J. E. Sheppard, Brooklyn, N. Y.
E. L. Shurly, Detroit, Mich.
William M. Sweet, Philadelphia.
Samuel Theobald, Baltimore, Md.
A. G. Thomson, Philadelphia.
Clarence A. Veasey, Philadelphia.
John E. Weeks, New York.
Casey A. Wood, Chicago, Ill.
Jonathan Wright, Brooklyn.
H. V. Würdemann, Milwaukee, Wis.

* **AN AMERICAN YEAR-BOOK OF MEDICINE AND SURGERY.** A Yearly Digest of Scientific Progress and Authoritative Opinion in all branches of Medicine and Surgery, drawn from journals, monographs, and text-books of the leading American and Foreign authors and investigators. Collected and arranged, with critical editorial comments, by eminent American specialists and teachers, under the general editorial charge of GEORGE M. GOULD, M.D. One handsome imperial octavo volume of about 1200 pages. Uniform in style, size, and general make-up with the "American Text-Book" Series. Cloth, $6.50 net; Half-Morocco, $7.50 net.

Now Ready, Volumes for 1896, 1897, 1898, 1899.

Notwithstanding the rapid multiplication of medical and surgical works, still these publications fail to meet fully the requirements of the *general physician*, inasmuch as he feels the need of something more than mere text-books of well-known principles of medical science.

This deficiency would best be met by current journalistic literature, but most practitioners have scant access to this almost unlimited source of information, and the busy practiser has but little time to search out in periodicals the many interesting cases whose study would doubtless be of inestimable value in his practice. Therefore, a work which places before the physician in convenient form *an epitomization of this literature by persons competent to pronounce upon*

The Value of a Discovery or of a Method of Treatment

cannot but command his highest appreciation. It is this critical and judicial function that is assumed by the Editorial staff of the "American Year-Book of Medicine and Surgery."

CONTRIBUTORS:

Dr. Samuel W. Abbott, Boston.	Dr. Howard E. Hansell, Philadelphia.
John J. Abel, Baltimore.	M. B. Hartzell, Philadelphia.
J. M. Baldy, Philadelphia.	Barton Cooke Hirst, Philadelphia.
Charles H. Burnett, Philadelphia.	E. Fletcher Ingals, Chicago.
Archibald Church, Chicago.	Wyatt Johnston, Montreal.
J. Chalmers DaCosta, Philadelphia.	W. W. Keen, Philadelphia.
W. A. N. Dorland, Philadelphia.	Henry G. Ohls, Chicago.
Louis A. Duhring, Philadelphia.	Wendell Reber, Philadelphia.
D. L. Edsall, Philadelphia.	David Riesman, Philadelphia.
Virgil P. Gibney, New York.	Louis Starr, Philadelphia.
Henry A. Griffin, New York.	Alfred Stengel, Philadelphia.
John Guitéras, Philadelphia.	G. N. Stewart, Cleveland.
C. A. Hamann, Cleveland.	J. R. Tillinghast, New York.
Alfred Hand, Jr., Philadelphia.	J. Hilton Waterman, New York.

"It is difficult to know which to admire most—the research and industry of the distinguished band of experts whom Dr. Gould has enlisted in the service of the Year-Book, or the wealth and abundance of the contributions to every department of science that have been deemed worthy of analysis. . . . It is much more than a mere compilation of abstracts, for, as each section is entrusted to experienced and able contributors, the reader has the advantage of certain critical commentaries and expositions . . . proceeding from writers fully qualified to perform these tasks. . . . It is emphatically a book which should find a place in every medical library, and is in several respects more useful than the famous 'Jahrbücher' of Germany."—*London Lancet.*

*ANOMALIES AND CURIOSITIES OF MEDICINE. By GEORGE M. GOULD, M.D., and WALTER L. PYLE, M.D. An encyclopedic collection of are and extraordinary cases and of the most striking instances of abnormality in all branches of Medicine and Surgery, derived from an exhaustive research of medical literature from its origin to the present day, abstracted, classified, annotated, and indexed. Handsome imperial octavo volume of 968 pages, with 295 engravings in the text, and 12 full-page plates. Cloth, $6.00 net; Half-Morocco, $7.00 net.

Several years of exhaustive research have been spent by the authors in the great medical libraries of the United States and Europe in collecting the material for this work. **Medical literature of all ages and all languages** has been carefully searched, as a glance at the Bibliographic Index will show. The facts, which will be of **extreme value to the author and lecturer**, have been arranged and annotated, and full reference footnotes given, indicating whence they have been obtained.

In view of the persistent and dominant interest in the anomalous and curious, a **thorough and systematic collection** of this kind (the first of which the authors have knowledge) must have its own peculiar sphere of usefulness.

As a complete and authoritative **Book of Reference** it will be of value not only to members of the medical profession, but to all persons interested in general scientific, sociologic, and medico-legal topics; in fact, the general interest of the subject and the dearth of any complete work upon it make this volume **one of the most important literary innovations of the day.**

"One of the most valuable contributions ever made to medical literature. It is, so far as we know, absolutely unique, and every page is as fascinating as a novel. Not alone for the medical profession has this volume value : it will serve as a book of reference for all who are interested in general scientific, sociologic, or medico-legal topics."—*Brooklyn Medical Journal.*

NERVOUS AND MENTAL DISEASES. By ARCHIBALD CHURCH, M. D., Professor of Clinical Neurology, Mental Diseases, and Medical Jurisprudence, Northwestern University Medical School; and FREDERICK PETERSON, M. D., Clinical Professor of Mental Diseases, Woman's Medical College, New York. Handsome octavo volume of 843 pages, with over 300 illustrations. Prices: Cloth, $5.00 net; Half-Morocco, $6.00 net.

Just Issued.

This book is intended to furnish students and practitioners with a practical, working knowledge of nervous and mental diseases. Written by men of wide experience and authority, it presents the many recent additions to the subject. The book is not filled with an extended dissertation on anatomy and pathology, but, treating these points in connection with special conditions, it lays particular stress on methods of examination, diagnosis, and treatment. In this respect the work is unusually complete and valuable, laying down the definite courses of procedure which the authors have found to be most generally satisfactory.

A TEXT-BOOK OF PATHOLOGY. By ALFRED STENGEL, M. D., Professor of Clinical Medicine in the University of Pennsylvania; Physician to the Philadelphia Hospital; Physician to the Children's Hospital, Philadelphia. Handsome octavo volume of 848 pages, with 362 illustrations, many of which are in colors. Prices: Cloth, $4.00 net; Half-Morocco, $5.00 net.

Second Edition.

In this work the practical application of pathological facts to clinical medicine is considered more fully than is customary in works on pathology. While the subject of pathology is treated in the broadest way consistent with the size of the book, an effort has been made to present the subject from the point of view of the clinician. The general relations of bacteriology to pathology are discussed at considerable length, as the importance of these branches deserves. It will be found that the recent knowledge is fully considered, as well as older and more widely-known facts.

" I consider the work abreast of modern pathology, and useful to both students and practitioners. It presents in a concise and well-considered form the essential facts of general and special pathological anatomy, with more than usual emphasis upon pathological physiology."
—WILLIAM H. WELCH, *Professor of Pathology, Johns Hopkins University, Baltimore, Md.*

" I regard it as the most serviceable text-book for students on this subject yet written by an American author."—L. HEKTOEN, *Professor of Pathology, Rush Medical College, Chicago, Ill.*

A TEXT-BOOK OF OBSTETRICS. By BARTON COOKE HIRST, M.D., Professor of Obstetrics in the University of Pennsylvania. Handsome octavo volume of 846 pages, with 618 illustrations and seven colored plates. Prices: Cloth, $5.00 net; Half-Morocco, $6.00 net.

This work, which has been in course of preparation for several years, is intended as an ideal text-book for the student no less than an advanced treatise for the obstetrician and for general practitioners. It represents the very latest teaching in the practice of obstetrics by a man of extended experience and recognized authority. The book emphasizes especially, as a work on obstetrics should, the practical side of the subject, and to this end presents an unusually large collection of illustrations. A great number of these are new and original, and the whole collection will form a complete atlas of obstetrical practice. An extremely valuable feature of the book is the large number of references to cases, authorities, sources, etc., forming, as it does, a valuable bibliography of the most recent and authoritative literature on the subject of obstetrics. As already stated, this work records the wide practical experience of the author, which fact, combined with the brilliant presentation of the subject, will doubtless render this one of the most notable books on obstetrics that has yet appeared.

" The illustrations are numerous and are works of art, many of them appearing for the first time. The arrangement of the subject-matter, the foot-notes, and index are beyond criticism. The author's style, though condensed, is singularly clear, so that it is never necessary to re-read a sentence in order to grasp its meaning. As a true model of what a modern text-book in obstetrics should be, we feel justified in affirming that Dr. Hirst's book is without a rival."—*New York Medical Record.*

A TEXT-BOOK OF THE PRACTICE OF MEDICINE. By James M. Anders, M.D., Ph.D., LL.D., Professor of the Practice of Medicine and of Clinical Medicine, Medico-Chirurgical College, Philadelphia. In one handsome octavo volume of 1287 pages, fully illustrated. Cloth, $5.50 net; Sheep or Half-Morocco, $6.50 net.

THIRD EDITION, THOROUGHLY REVISED.

This work gives in a comprehensive manner the results of the latest scientific studies bearing upon medical affections, and portrays with rare force and clearness the clinical pictures of the different diseases considered. The practical points, particularly with reference to diagnosis and treatment, are completely stated and are presented in a most convenient form; for example, the differential diagnosis has in many instances been tabulated, no less than fifty-six diagnostic tables being given.
The first edition of this work having been exhausted in so short a time, the author has not found it necessary to make an extensive revision, but has simply availed himself of the opportunity to make a few changes of minor importance.

"It is an excellent book—concise, comprehensive, thorough, and up to date. It is a credit to you; but, more than that, it is a credit to the profession of Philadelphia—to us."
—James C. Wilson, *Professor of the Practice of Medicine and Clinical Medicine, Jefferson Medical College, Philadelphia.*

"I consider Dr. Anders' book not only the best late work on Medical Practice, but by far the best that has ever been published. It is concise, systematic, thorough, and fully up to date in everything. I consider it a great credit to both the author and the publisher."—A. C. Cowperthwaite, *President of the Illinois Homeopathic Medical Association.*

DISEASES OF THE STOMACH. By William W. Van Valzah, M.D., Professor of General Medicine and Diseases of the Digestive System and the Blood, New York Polyclinic; and J. Douglas Nisbet, M.D., Adjunct Professor of General Medicine and Diseases of the Digestive System and the Blood, New York Polyclinic. Octavo volume of 674 pages, illustrated. Cloth, $3.50 net.

An eminently practical book, intended as a guide to the student, an aid to the physician, and a contribution to scientific medicine. It aims to give a complete description of the modern methods of diagnosis and treatment of diseases of the stomach, and to reconstruct the pathology of the stomach in keeping with the revelations of scientific research. The book is clear, practical, and complete, and contains the results of the authors' investigations and of their extensive experience as specialists. Particular attention is given to the important subject of dietetic treatment. The diet-lists are very complete, and are so arranged that selections can readily be made to suit individual cases.

"This is the most satisfactory work on the subject in the English language."—*Chicago Medical Recorder.*

"The article on diet and general medication is one of the most valuable in the book, and should be read by every practising physician."—*New York Medical Journal.*

SURGICAL DIAGNOSIS AND TREATMENT. By J. W. MAC-
DONALD, M. D., Edin., F. R. C. S., Edin., Professor of the Practice of Surgery and of Clinical Surgery in Hamline University; Visiting Surgeon to St. Barnabas' Hospital, Minneapolis, etc. Handsome octavo volume of 800 pages, profusely illustrated. Cloth, $5.00 net; Half-Morocco, $6.00 net.

This work aims in a comprehensive manner to furnish a guide in matters of surgical diagnosis. It sets forth in a systematic way the necessities of examinations and the proper methods of making them. The various portions of the body are then taken up in order and the diseases and injuries thereof succinctly considered and the treatment briefly indicated. Practically all the modern and approved operations are described with thoroughness and clearness. The work concludes with a chapter on the use of the Röntgen rays in surgery.

"The work is brimful of just the kind of practical information that is useful alike to students and practitioners. It is a pleasure to commend the book because of its intrinsic value to the medical practitioner."—*Cincinnati Lancet-Clinic.*

PATHOLOGICAL TECHNIQUE. A Practical Manual for Laboratory Work in Pathology, Bacteriology, and Morbid Anatomy, with chapters on Post-Mortem Technique and the Performance of Autopsies. By FRANK B. MALLORY, A. M., M. D., Assistant Professor of Pathology, Harvard University Medical School, Boston; and JAMES H. WRIGHT, A. M., M. D., Instructor in Pathology, Harvard University Medical School, Boston. Octavo volume of 396 pages, handsomely illustrated. Cloth, $2.50 net.

This book is designed especially for practical use in pathological laboratories, both as a guide to beginners and as a source of reference for the advanced. The book will also meet the wants of practitioners who have opportunity to do general pathological work. Besides the methods of post-mortem examinations and of bacteriological and histological investigations connected with autopsies, the special methods employed in clinical bacteriology and pathology have been fully discussed.

"One of the most complete works on the subject, and one which should be in the library of every physician who hopes to keep pace with the great advances made in pathology."—*Journal of American Medical Association.*

THE SURGICAL COMPLICATIONS AND SEQUELS OF TYPHOID FEVER. By WM. W. KEEN, M. D., LL.D., Professor of the Principles of Surgery and of Clinical Surgery, Jefferson Medical College, Philadelphia. Octavo volume of 386 pages, illustrated. Cloth, $3.00 net.

This monograph is the only one in any language covering the entire subject of the Surgical Complications and Sequels of Typhoid Fever. The work will prove to be of importance and interest not only to the general surgeon and physician, but also to many specialists—laryngologists, ophthalmologists, gynecologists, pathologists, and bacteriologists—as the subject has an important bearing upon each one of their spheres. The author's conclusions are based on reports of over 1700 cases, including practically all those recorded in the last fifty years. Reports of cases have been brought down to date, many having been added while the work was in press.

"This is probably the first and only work in the English language that gives the reader a clear view of what typhoid fever really is, and what it does and can do to the human organism. This book should be in the possession of every medical man in America."—*American Medico-Surgical Bulletin.*

MODERN SURGERY, GENERAL AND OPERATIVE. By JOHN CHALMERS DACOSTA, M.D., Clinical Professor of Surgery, Jefferson Medical College, Philadelphia; Surgeon to the Philadelphia Hospital, etc. Handsome octavo volume of 911 pages, profusely illustrated. Cloth, $4.00 net; Half-Morocco, $5.00 net.

Second Edition, Rewritten and Greatly Enlarged.

The remarkable success attending DaCosta's Manual of Surgery, and the general favor with which it has been received, have led the author in this revision to produce a complete treatise on modern surgery along the same lines that made the former edition so successful. The book has been entirely rewritten and very much enlarged. The old edition has long been a favorite not only with students and teachers, but also with practising physicians and surgeons, and it is believed that the present work will find an even wider field of usefulness.

"We know of no small work on surgery in the English language which so well fulfils the requirements of the modern student."—*Medico-Chirurgical Journal*, Bristol, England.

"The author has presented concisely and accurately the principles of modern surgery. The book is a valuable one which can be recommended to students and is of great value to the general practitioner."—*American Journal of the Medical Sciences*.

A MANUAL OF ORTHOPEDIC SURGERY. By JAMES E. MOORE, M.D., Professor of Orthopedics and Adjunct Professor of Clinical Surgery, University of Minnesota, College of Medicine and Surgery. Octavo volume of 356 pages, with 177 beautiful illustrations from photographs made specially for this work. Cloth, $2.50 net.

A practical book based upon the author's experience, in which special stress is laid upon early diagnosis and treatment such as can be carried out by the general practitioner. The teachings of the author are in accordance with his belief that true conservatism is to be found in the middle course between the surgeon who operates too frequently and the orthopedist who seldom operates.

"A very demonstrative work, every illustration of which conveys a lesson. The work is a most excellent and commendable one, which we can certainly endorse with pleasure."—*St. Louis Medical and Surgical Journal*.

ELEMENTARY BANDAGING AND SURGICAL DRESSING. With Directions concerning the Immediate Treatment of Cases of Emergency. For the use of Dressers and Nurses. By WALTER PYE, F.R.C.S., late Surgeon to St. Mary's Hospital, London. Small 12mo, with over 80 illustrations. Cloth, flexible covers, 75 cents net.

This little book is chiefly a condensation of those portions of Pye's "Surgical Handicraft" which deal with bandaging, splinting, etc., and of those which treat of the management in the first instance of cases of emergency. The directions given are thoroughly practical, and the book will prove extremely useful to students, surgical nurses, and dressers.

"The author writes well, the diagrams are clear, and the book itself is small and portable, although the paper and type are good."—*British Medical Journal*.

A TEXT-BOOK OF MATERIA MEDICA, THERAPEUTICS AND PHARMACOLOGY. By GEORGE F. BUTLER, PH.G., M.D., Professor of Materia Medica and of Clinical Medicine in the College of Physicians and Surgeons, Chicago; Professor of Materia Medica and Therapeutics, Northwestern University, Woman's Medical School, etc. Octavo, 860 pages, illustrated. Cloth, $4.00 net; Sheep, $5.00 net.

Third Edition, Thoroughly Revised.

A clear, concise, and practical text-book, adapted for permanent reference no less than for the requirements of the class-room.

The recent important additions made to our knowledge of the physiological action of drugs are fully discussed in the present edition. Many alterations also have been made in the chapters on Diuretics and Cathartics.

"Taken as a whole, the book may fairly be considered as one of the most satisfactory of any single-volume works on materia medica in the market."—*Journal of the American Medical Association.*

TUBERCULOSIS OF THE GENITO-URINARY ORGANS, MALE AND FEMALE. By NICHOLAS SENN, M.D., PH.D., LL.D., Professor of the Practice of Surgery and of Clinical Surgery, Rush Medical College, Chicago. Handsome octavo volume of 320 pages, illustrated. Cloth, $3.00 net.

Tuberculosis of the male and female genito-urinary organs is such a frequent, distressing, and fatal affection that a special treatise on the subject appears to fill a gap in medical literature. In the present work the bacteriology of the subject has received due attention, the modern resources employed in the differential diagnosis between tubercular and other inflammatory affections are fully described, and the medical and surgical therapeutics are discussed in detail.

"An important book upon an important subject, and written by a man of mature judgment and wide experience. The author has given us an instructive book upon one of the most important subjects of the day."—*Clinical Reporter.*

"A work which adds another to the many obligations the profession owes the talented author."—*Chicago Medical Recorder.*

A TEXT-BOOK OF DISEASES OF WOMEN. By CHARLES B. PENROSE, M.D., PH.D., Professor of Gynecology in the University of Pennsylvania; Surgeon to the Gynecean Hospital, Philadelphia. Octavo volume of 529 pages, with 217 illustrations, nearly all from drawings made for this work. Cloth, $3.50 net.

Second Edition, Revised.

In this work, which has been written for both the student of gynecology and the general practitioner, the author presents the best teaching of modern gynecology untrammelled by antiquated theories or methods of treatment. In most instances but one plan of treatment is recommended, to avoid confusing the student or the physician who consults the book for practical guidance.

"I shall value very highly the copy of Penrose's 'Diseases of Women' received. I have already recommended it to my class as THE BEST book."—HOWARD A. KELLY, *Professor of Gynecology and Obstetrics, Johns Hopkins University, Baltimore, Md.*

"The book is to be commended without reserve, not only to the student but to the general practitioner who wishes to have the latest and best modes of treatment explained with absolute clearness."—*Therapeutic Gazette.*

SURGICAL PATHOLOGY AND THERAPEUTICS. By JOHN COLLINS WARREN, M. D., LL.D., Professor of Surgery, Medical Department Harvard University; Surgeon to the Massachusetts General Hospital, etc. A handsome octavo volume of 832 pages, with 136 relief and lithographic illustrations, 33 of which are printed in colors, and all of which were drawn by William J. Kaula from original specimens. Prices: Cloth, $6.00 net; Half-Morocco, $7.00 net.

Without Exception, the Illustrations are the Best ever Seen in a Work of this Kind.

"A most striking and very excellent feature of this book is its illustrations. Without exception, from the point of accuracy and artistic merit, they are the best ever seen in a work of this kind. * * * Many of those representing microscopic pictures are so perfect in their coloring and detail as almost to give the beholder the impression that he is looking down the barrel of a microscope at a well-mounted section."—*Annals of Surgery*, Philadelphia.

"It is the handsomest specimen of book-making * * * that has ever been issued from the American medical press."—*American Journal of the Medical Sciences*, Philadelphia.

PATHOLOGY AND SURGICAL TREATMENT OF TUMORS. By N. SENN, M. D., Ph. D., LL. D., Professor of Practice of Surgery and of Clinical Surgery, Rush Medical College; Professor of Surgery, Chicago Polyclinic; Attending Surgeon to Presbyterian Hospital; Surgeon-in-Chief, St. Joseph's Hospital, Chicago. One volume of 710 pages, with 515 engravings, including full-page colored plates. Prices: Cloth, $6.00 net; Half-Morocco, $7.00 net.

Books specially devoted to this subject are few, and in our text-books and systems of surgery this part of surgical pathology is usually condensed to a degree incompatible with its scientific and clinical importance. The author spent many years in collecting the material for this work, and has taken great pains to present it in a manner that should prove useful as a text-book for the student, a work of reference for the practitioner, and a reliable guide for the surgeon.

"The most exhaustive of any recent book in English on this subject. It is well illustrated, and will doubtless remain as the principal monograph on the subject in our language for some years. The book is handsomely illustrated and printed, ... and the author has given a notable and lasting contribution to surgery."—*Journal of the American Medical Association*, Chicago.

LECTURES ON RENAL AND URINARY DISEASES. By ROBERT SAUNDBY, M. D., Edin., Fellow of the Royal College of Physicians, London, and of the Royal Medico-Chirurgical Society; Physician to the General Hospital. Octavo volume of 434 pages, with numerous illustrations and 4 colored plates. Cloth, $2.50 net.

"The volume makes a favorable impression at once. The style is clear and succinct. We cannot find any part of the subject in which the views expressed are not carefully thought out and fortified by evidence drawn from the most recent sources. The book may be cordially recommended."—*British Medical Journal*.

"The work represents the present knowledge of renal and urinary diseases. It is admirably written and is accurately scientific."—*Medical News*.

A NEW PRONOUNCING DICTIONARY OF MEDICINE, with Phonetic Pronunciation, Accentuation, Etymology, etc. By JOHN M. KEATING, M. D., LL.D., Fellow of the College of Physicians of Philadelphia; Vice-President of the American Pædiatric Society; Ex-President of the Association of Life Insurance Medical Directors; Editor "Cyclopædia of the Diseases of Children," etc.; and HENRY HAMILTON, author of "A New Translation of Virgil's Æneid into English Rhyme;" co-author of "Saunders' Medical Lexicon," etc.; with the Collaboration of J. CHALMERS DACOSTA, M. D., and FREDERICK A. PACKARD, M. D. With an Appendix containing important Tables of Bacilli, Micrococci, Leucomaïnes, Ptomaïnes, Drugs and Materials used in Antiseptic Surgery, Poisons and their Antidotes, Weights and Measures, Thermometric Scales, New Official and Unofficial Drugs, etc. One very attractive volume of over 800 pages. Second Revised Edition. Prices: Cloth, $5.00 net; Sheep or Half-Morocco, $6.00 net; with Denison's Patent Ready-Reference Index; without patent index, Cloth, $4.00 net; Sheep or Half-Morocco, $5.00 net.

PROFESSIONAL OPINIONS.

"I am much pleased with Keating's Dictionary, and shall take pleasure in recommending it to my classes."
HENRY M. LYMAN, M. D.,
Professor of Principles and Practice of Medicine, Rush Medical College, Chicago, Ill.

"I am convinced that it will be a very valuable adjunct to my study-table, convenient in size and sufficiently full for ordinary use."
C. A. LINDSLEY, M. D.,
Professor of Theory and Practice of Medicine, Medical Dept. Yale University; Secretary Connecticut State Board of Health, New Haven, Conn.

AUTOBIOGRAPHY OF SAMUEL D. GROSS, M. D., Emeritus Professor of Surgery in the Jefferson Medical College of Philadelphia, with Reminiscences of His Times and Contemporaries. Edited by his sons, SAMUEL W. GROSS, M. D., LL.D., late Professor of Principles of Surgery and of Clinical Surgery in the Jefferson Medical College, and A. HALLER GROSS, A. M., of the Philadelphia Bar. Preceded by a Memoir of Dr. Gross, by the late Austin Flint, M. D., LL.D. In two handsome volumes, each containing over 400 pages, demy 8vo, extra cloth, gilt tops, with fine Frontispiece engraved on steel. Price per Volume, $2.50 net.

This autobiography, which was continued by the late eminent surgeon until within three months of his death, contains a full and accurate history of his early struggles, trials, and subsequent successes, told in a singularly interesting and charming manner, and embraces short and graphic pen-portraits of many of the most distinguished men—surgeons, physicians, divines, lawyers, statesmen, scientists, etc.—with whom he was brought in contact in America and in Europe; the whole forming a retrospect of more than three-quarters of a century.

PRACTICAL POINTS IN NURSING. For Nurses in Private Practice. By EMILY A. M. STONEY, Graduate of the Training-School for Nurses, Lawrence, Mass.; Superintendent of the Training-School for Nurses, Carney Hospital, South Boston, Mass. 456 pages, handsomely illustrated with 73 engravings in the text, and 9 colored and half-tone plates. Cloth. Price, $1.75 net.

SECOND EDITION, THOROUGHLY REVISED.

In this volume the author explains, in popular language and in the shortest possible form, the entire range of *private* nursing as distinguished from *hospital* nursing, and the nurse is instructed how best to meet the various emergencies of medical and surgical cases when distant from medical or surgical aid or when thrown on her own resources.

An especially valuable feature of the work will be found in the directions to the nurse how to *improvise* everything ordinarily needed in the sick-room, where the embarrassment of the nurse, owing to the want of proper appliances, is frequently extreme.

The work has been logically divided into the following sections:

I. The Nurse: her responsibilities, qualifications, equipment, etc.
II. The Sick-Room: its selection, preparation, and management.
III. The Patient: duties of the nurse in medical, surgical, obstetric, and gynecologic cases.
IV. Nursing in Accidents and Emergencies.
V. Nursing in Special Medical Cases.
VI. Nursing of the New-born and Sick Children.
VII. Physiology and Descriptive Anatomy.

The APPENDIX contains much information in compact form that will be found of great value to the nurse, including Rules for Feeding the Sick; Recipes for Invalid Foods and Beverages; Tables of Weights and Measures; Table for Computing the Date of Labor; List of Abbreviations; Dose-List; and a full and complete Glossary of Medical Terms and Nursing Treatment.

"This is a well-written, eminently practical volume, which covers the entire range of private nursing as distinguished from hospital nursing, and instructs the nurse how best to meet the various emergencies which may arise and how to prepare everything ordinarily needed in the illness of her patient."—*American Journal of Obstetrics and Diseases of Women and Children*, Aug., 1896.

A TEXT-BOOK OF BACTERIOLOGY, including the Etiology and Prevention of Infective Diseases and an account of Yeasts and Moulds, Hæmatozoa, and Psorosperms. By EDGAR M. CROOKSHANK, M. B., Professor of Comparative Pathology and Bacteriology, King's College, London. A handsome octavo volume of 700 pages, with 273 engravings in the text, and 22 original and colored plates. Price, $6.50 net.

This book, though nominally a Fourth Edition of Professor Crookshank's "MANUAL OF BACTERIOLOGY," is practically a new work, the old one having been reconstructed, greatly enlarged, revised throughout, and largely rewritten, forming a text-book for the Bacteriological Laboratory, for Medical Officers of Health, and for Veterinary Inspectors.

MEDICAL DIAGNOSIS. By Dr. OSWALD VIERORDT, Professor of Medicine at the University of Heidelberg. Translated, with additions, from the Fifth Enlarged German Edition, with the author's permission, by FRANCIS H. STUART, A. M., M. D. In one handsome royal-octavo volume of 600 pages. 194 fine wood-cuts in the text, many of them in colors. Prices: Cloth, $4.00 net; Sheep or Half-Morocco, $5.00 net.

FOURTH AMERICAN EDITION, FROM THE FIFTH REVISED AND ENLARGED GERMAN EDITION.

In this work, as in no other hitherto published, are given full and accurate explanations of the phenomena observed at the bedside. It is distinctly a clinical work by a master teacher, characterized by thoroughness, fulness, and accuracy. It is a mine of information upon the points that are so often passed over without explanation. Especial attention has been given to the germ-theory as a factor in the origin of disease.

The present edition of this highly successful work has been translated from the fifth German edition. Many alterations have been made throughout the book, but especially in the sections on Gastric Digestion and the Nervous System.

It will be found that all the qualities which served to make the earlier editions so acceptable have been developed with the evolution of the work to its present form.

THE PICTORIAL ATLAS OF SKIN DISEASES AND SYPHILITIC AFFECTIONS. (American Edition.) Translation from the French. Edited by J. J. PRINGLE, M. B., F. R. C. P., Assistant Physician to, and Physician to the department for Diseases of the Skin at, the Middlesex Hospital, London. Photo-lithochromes from the famous models of dermatological and syphilitic cases in the Museum of the Saint-Louis Hospital, Paris, with explanatory wood-cuts and letter-press. In 12 Parts, at $3.00 per Part.

"Of all the atlases of skin diseases which have been published in recent years, the present one promises to be of greatest interest and value, especially from the standpoint of the general practitioner."—*American Medico-Surgical Bulletin*, Feb. 22, 1896.

"The introduction of explanatory wood-cuts in the text is a novel and most important feature which greatly furthers the easier understanding of the excellent plates, than which nothing, we venture to say, has been seen better in point of correctness, beauty, and general merit."—*New York Medical Journal*, Feb. 15, 1896.

"An interesting feature of the Atlas is the descriptive text, which is written for each picture by the physician who treated the case or at whose instigation the models have been made. We predict for this truly beautiful work a large circulation in all parts of the medical world where the names *St. Louis* and *Baretta* have preceded it."—*Medical Record*, N. Y., Feb. 1, 1896.

A TEXT-BOOK OF MECHANO-THERAPY (MASSAGE AND MEDICAL GYMNASTICS). By AXEL V. GRAFSTROM, B. Sc., M. D., late Lieutenant in the Royal Swedish Army; late House Physician, City Hospital, Blackwell's Island, New York. 12mo, 139 pages, illustrated. Cloth, $1.00 net.

DISEASES OF THE EYE. A Hand-Book of Ophthalmic Practice. By G. E. DE SCHWEINITZ, M. D., Professor of Ophthalmology in the Jefferson Medical College, Philadelphia, etc. A handsome royal-octavo volume of 696 pages, with 255 fine illustrations, many of which are original, and 2 chromo-lithographic plates. Prices: Cloth, $4.00 net; Sheep or Half-Morocco, $5.00 net.

THIRD EDITION, THOROUGHLY REVISED.

In the third edition of this text-book, destined, it is hoped, to meet the favorable reception which has been accorded to its predecessors, the work has been revised thoroughly, and much new matter has been introduced. Particular attention has been given to the important relations which micro-organisms bear to many ocular diseases. A number of special paragraphs on new subjects have been introduced, and certain articles, including a portion of the chapter on Operations, have been largely rewritten, or at least materially changed. A number of new illustrations have been added. The Appendix contains a full description of the method of determining the corneal astigmatism with the ophthalmometer of Javal and Schiötz, and the rotation of the eyes with the tropometer of Stevens.

"A work that will meet the requirements not only of the specialist, but of the general practitioner in a rare degree. I am satisfied that unusual success awaits it."
WILLIAM PEPPER, M. D.
Provost and Professor of Theory and Practice of Medicine and Clinical Medicine in the University of Pennsylvania.

"A clearly written, comprehensive manual. . . . One which we can commend to students as a reliable text-book, written with an evident knowledge of the wants of those entering upon the study of this special branch of medical science."—*British Medical Journal.*

"It is hardly too much to say that for the student and practitioner beginning the study of Ophthalmology, it is the best single volume at present published."—*Medical News.*

"It is a very useful, satisfactory, and safe guide for the student and the practitioner, and one of the best works of this scope in the English language."—*Annals of Ophthalmology.*

DISEASES OF WOMEN. By J. BLAND SUTTON, F. R. C. S., Assistant Surgeon to Middlesex Hospital, and Surgeon to Chelsea Hospital, London; and ARTHUR E. GILES, M. D., B. Sc., Lond., F. R. C. S., Edin., Assistant Surgeon to Chelsea Hospital, London. 436 pages, handsomely illustrated. Cloth, $2.50 net.

The authors have placed in the hands of the physician and student a concise yet comprehensive guide to the study of gynecology in its most modern development. It has been their aim to relate facts and describe methods belonging to the science and art of gynecology in a way that will prove useful to students for examination purposes, and which will also enable the general physician to practice this important department of surgery with advantage to his patients and with satisfaction to himself.

"The book is very well prepared, and is certain to be well received by the medical public."
—*British Medical Journal.*

"The text has been carefully prepared. Nothing essential has been omitted, and its teachings are those recommended by the leading authorities of the day."—*Journal of the American Medical Association.*

TEXT-BOOK UPON THE PATHOGENIC BACTERIA. Specially written for Students of Medicine. By JOSEPH MCFARLAND, M. D., Professor of Pathology and Bacteriology in the Medico-Chirurgical College of Philadelphia, etc. 497 pages, finely illustrated. Price, Cloth, $2.50 net.

SECOND EDITION, REVISED AND GREATLY ENLARGED.

The work is intended to be a text-book for the medical student and for the practitioner who has had no recent laboratory training in this department of medical science. The instructions given as to needed apparatus, cultures, stainings, microscopic examinations, etc. are ample for the student's needs, and will afford to the physician much information that will interest and profit him relative to a subject which modern science shows to go far in explaining the etiology of many diseased conditions.

In this second edition the work has been brought up to date in all departments of the subject, and numerous additions have been made to the technique in the endeavor to make the book fulfil the double purpose of a systematic work upon bacteria and a laboratory guide.

"It is excellently adapted for the medical students and practitioners for whom it is avowedly written. . . . The descriptions given are accurate and readable, and the book should prove useful to those for whom it is written.—*London Lancet*, Aug. 29, 1896.

"The author has succeded admirably in presenting the essential details of bacteriological technics, together with a judiciously chosen summary of our present knowledge of pathogenic bacteria. . . . The work, we think, should have a wide circulation among English-speaking students of medicine."—*N. Y. Medical Journal*, April 4, 1896.

"The book will be found of considerable use by medical men who have not had a special bacteriological training, and who desire to understand this important branch of medical science."—*Edinburgh Medical Journal*, July, 1896.

LABORATORY GUIDE FOR THE BACTERIOLOGIST. By LANGDON FROTHINGHAM, M. D. V., Assistant in Bacteriology and Veterinary Science, Sheffield Scientific School, Yale University. Illustrated. Price, Cloth, 75 cents.

The technical methods involved in bacteria-culture, methods of staining, and microscopical study are fully described and arranged as simply and concisely as possible. The book is especially intended for use in laboratory work

"It is a convenient and useful little work, and will more than repay the outlay necessary for its purchase in the saving of time which would otherwise be consumed in looking up the various points of technique so clearly and concisely laid down in its pages."—*American Med.-Surg. Bulletin*.

FEEDING IN EARLY INFANCY. By ARTHUR V. MEIGS, M. D. Bound in limp cloth, flush edges. Price, 25 cents net.

SYNOPSIS: Analyses of Milk—Importance of the Subject of Feeding in Early Infancy—Proportion of Casein and Sugar in Human Milk—Time to Begin Artificial Feeding of Infants—Amount of Food to be Administered at Each Feeding—Intervals between Feedings—Increase in Amount of Food at Different Periods of Infant Development—Unsuitableness of Condensed Milk as a Substitute for Mother's Milk—Objections to Sterilization or "Pasteurization" of Milk—Advances made in the Method of Artificial Feeding of Infants.

MATERIA MEDICA FOR NURSES. By EMILY A. M. STONEY, Graduate of the Training-school for Nurses, Lawrence, Mass.; late Superintendent of the Training-school for Nurses, Carney Hospital, South Boston, Mass. Handsome octavo, 300 pages. Cloth, $1.50 net.

The present book differs from other similar works in several features, all of which are introduced to render it more practical and generally useful. The general plan of contents follows the lines laid down in training-schools for nurses, but the book contains much useful matter not usually included in works of this character, such as Poison-emergencies, Ready Dose-list, Weights and Measures, etc., as well as a Glossary, defining all the terms in Materia Medica, and describing all the latest drugs and remedies, which have been generally neglected by other books of the kind.

ESSENTIALS OF ANATOMY AND MANUAL OF PRACTICAL DISSECTION, containing "Hints on Dissection" By CHARLES B. NANCREDE, M. D., Professor of Surgery and Clinical Surgery in the University of Michigan, Ann Arbor; Corresponding Member of the Royal Academy of Medicine, Rome, Italy; late Surgeon Jefferson Medical College, etc. Fourth and revised edition. Post 8vo, over 500 pages, with handsome full-page lithographic plates in colors, and over 200 illustrations. Price: Extra Cloth or Oilcloth for the dissection-room, $2.00 net.

Neither pains nor expense has been spared to make this work the most exhaustive yet concise Student's Manual of Anatomy and Dissection ever published, either in America or in Europe.

The colored plates are designed to aid the student in dissecting the muscles arteries, veins, and nerves. The wood-cuts have all been specially drawn and engraved, and an Appendix added containing 60 illustrations representing the structure of the entire human skeleton, the whole being based on the eleventh edition of Gray's *Anatomy*.

A MANUAL OF PRACTICE OF MEDICINE. By A. A. STEVENS, A. M., M. D., Instructor in Physical Diagnosis in the University of Pennsylvania, and Professor of Pathology in the Woman's Medical College of Pennsylvania. Specially intended for students preparing for graduation and hospital examinations. Post 8vo, 519 pages. Numerous illustrations and selected formulæ. Price, bound in flexible leather, $2.00 net.

FIFTH EDITION, REVISED AND ENLARGED.

Contributions to the science of medicine have poured in so rapidly during the last quarter of a century that it is well-nigh impossible for the student, with the limited time at his disposal, to master elaborate treatises or to cull from them that knowledge which is absolutely essential. From an extended experience in teaching, the author has been enabled, by classification, to group allied symptoms, and by the judicious elimination of theories and redundant explanations to bring within a comparatively small compass a complete outline of the practice of medicine.

MANUAL OF MATERIA MEDICA AND THERAPEUTICS.
By A. A. STEVENS, A. M., M. D., Instructor of Physical Diagnosis in the University of Pennsylvania, and Professor of Pathology in the Woman's Medical College of Pennsylvania. 445 pages. Price, bound in flexible leather, $2.25.

SECOND EDITION, REVISED.

This wholly new volume, which is based on the last edition of the *Pharmacopœia*, comprehends the following sections: Physiological Action of Drugs; Drugs; Remedial Measures other than Drugs; Applied Therapeutics;. Incompatibility in Prescriptions; Table of Doses; Index of Drugs; and Index of Diseases; the treatment being elucidated by more than two hundred formulæ.

"The author is to be congratulated upon having presented the medical student with as accurate a manual of therapeutics as it is possible to prepare."—*Therapeutic Gazette.*

"Far superior to most of its class; in fact, it is very good. Moreover, the book is reliable and accurate."—*New York Medical Journal.*

"The author has faithfully presented modern therapeutics in a comprehensive work, . . . and it will be found a reliable guide."—*University Medical Magazine.*

NOTES ON THE NEWER REMEDIES: their Therapeutic Applications and Modes of Administration. By DAVID CERNA, M. D., PH. D., Demonstrator of and Lecturer on Experimental Therapeutics in the University of Pennsylvania. Post-octavo, 253 pages. Price, $1.25.

SECOND EDITION, RE-WRITTEN AND GREATLY ENLARGED.

The work takes up in alphabetical order all the newer remedies, giving their physical properties, solubility, therapeutic applications, administration, and chemical formula.

It thus forms a very valuable addition to the various works on therapeutics now in existence.

Chemists are so multiplying compounds, that, if each compound is to be thoroughly studied, investigations must be carried far enough to determine the practical importance of the new agents.

"Especially valuable because of its completeness, its accuracy, its systematic consideration of the properties and therapy of many remedies of which doctors generally know but little, expressed in a brief yet terse manner."—*Chicago Clinical Review.*

TEMPERATURE CHART. Prepared by D. T. LAINÉ, M. D. Size 8 x 13½ inches. Price, per pad of 25 charts, 50 cents.

A conveniently arranged chart for recording Temperature, with columns for daily amounts of Urinary and Fecal Excretions, Food, Remarks, etc. On the back of each chart is given in full the method of Brand in the treatment of Typhoid Fever.

A TEXT-BOOK OF HISTOLOGY, DESCRIPTIVE AND PRACTICAL. For the Use of Students. By ARTHUR CLARKSON, M. B., C. M., Edin., formerly Demonstrator of Physiology in the Owen's College, Manchester; late Demonstrator of Physiology in the Yorkshire College, Leeds. Large 8vo, 554 pages, with 22 engravings in the text, and 174 beautifully colored original illustrations. Price, strongly bound in Cloth, $6.00 net.

The purpose of the writer in this work has been to furnish the student of Histology, in one volume, with both the descriptive and the practical part of the science. The first two chapters are devoted to the consideration of the general methods of Histology; subsequently, in each chapter, the structure of the tissue or organ is first systematically described, the student is then taken tutorially over the specimens illustrating it, and, finally, an appendix affords a short note of the methods of preparation.

"The work must be considered a valuable addition to the list of available text-books, and is to be highly recommended."—*New York Medical Journal.*

"One of the best works for students we have ever noticed. We predict that the book will attain a well-deserved popularity among our students."—*Chicago Medical Recorder.*

THE PATHOLOGY AND TREATMENT OF SEXUAL IMPOTENCE. By VICTOR G. VECKI, M. D. From the second German edition, revised and rewritten. Demi-octavo, about 300 pages. Cloth, $2.00 net.

The subject of impotence has but seldom been treated in this country in the truly scientific spirit that it deserves, and this volume will come to many as a revelation of the possibilities of therapeusis in this important field. Dr. Vecki's work has long been favorably known, and the German book has received the highest consideration. This edition is more than a mere translation, for, although based on the German edition, it has been entirely rewritten by the author in English.

"The work can be recommended as a scholarly treatise on its subject, and it can be read with advantage by many practitioners."—*Journal of the American Medical Association.*

ARCHIVES OF CLINICAL SKIAGRAPHY. By SYDNEY ROWLAND, B. A., Camb. A series of collotype illustrations, with descriptive text, illustrating the applications of the New Photography to Medicine and Surgery. Price, per Part, $1.00. Parts I. to V. now ready.

The object of this publication is to put on record in permanent form some of the most striking applications of the new photography to the needs of Medicine and Surgery.

The progress of this new art has been so rapid that, although Prof. Röntgen's discovery is only a thing of yesterday, it has already taken its place among the approved and accepted aids to diagnosis.

DISEASES OF WOMEN. By HENRY J. GARRIGUES, A.M., M.D., Professor of Gynecology in the New York School of Clinical Medicine; Gynecologist to St. Mark's Hospital and to the German Dispensary, New York City. In one handsome octavo volume of 728 pages, illustrated by 335 engravings and colored plates. Prices: Cloth, $4.00 net; Sheep or Half-Morocco, $5.00 net.

A PRACTICAL work on gynecology for the use of students and practitioners, written in a terse and concise manner. The importance of a thorough knowledge of the anatomy of the female pelvic organs has been fully recognized by the author, and considerable space has been devoted to the subject. The chapters on Operations and on Treatment are thoroughly modern, and are based upon the large hospital and private practice of the author. The text is elucidated by a large number of illustrations and colored plates, many of them being original, and forming a complete atlas for studying *embryology* and the *anatomy* of the *female genitalia*, besides exemplifying, whenever needed, morbid conditions, instruments, apparatus, and operations.

Second Edition, Thoroughly Revised.

The first edition of this work met with a most appreciative reception by the medical press and profession both in this country and abroad, and was adopted as a text-book or recommended as a book of reference by nearly *one hundred* colleges in the United States and Canada. The author has availed himself of the opportunity afforded by this revision to embody the latest approved advances in the treatment employed in this important branch of Medicine. He has also more extensively expressed his own opinion on the comparative value of the different methods of treatment employed.

"One of the best text-books for students and practitioners which has been published in the English language; it is condensed, clear, and comprehensive. The profound learning and great clinical experience of the distinguished author find expression in this book in a most attractive and instructive form. Young practitioners, to whom experienced consultants may not be available, will find in this book invaluable counsel and help."

THAD. A. REAMY, M.D., LL.D.,
Professor of Clinical Gynecology, Medical College of Ohio; Gynecologist to the Good Samaritan and Cincinnati Hospitals.

A SYLLABUS OF GYNECOLOGY, arranged in conformity with "An American Text-Book of Gynecology." By J. W. LONG, M.D., Professor of Diseases of Women and Children, Medical College of Virginia, etc. Price, Cloth (interleaved), $1.00 net.

Based upon the teaching and methods laid down in the larger work, this will not only be useful as a supplementary volume, but to those who do not already possess the text-book it will also have an independent value as an aid to the practitioner in gynecological work, and to the student as a guide in the lecture-room, as the subject is presented in a manner at once systematic, clear, succinct, and practical.

THE AMERICAN POCKET MEDICAL DICTIONARY. Edited by W. A. NEWMAN DORLAND, M. D., Assistant Obstetrician to the Hospital of the University of Pennsylvania; Fellow of the American Academy of Medicine. Containing the pronunciation and definition of over 26,000 words used in medicine and the kindred sciences, with 64 extensive tables. Handsomely bound in flexible leather, limp, with gold edges and patent thumb index. Price, $1.25 net.

SECOND EDITION, REVISED.
Over 26,000 Words, 64 Valuable Tables.

This is the ideal pocket lexicon. It is an absolutely new book, and not a revision of any old work. It is complete, defining all the terms of modern medicine and forming a vocabulary of over 26,000 words. It gives the pronunciation of all the terms. It makes a special feature of the newer words neglected by other dictionaries. It contains a wealth of anatomical tables of special value to students. It forms a handy volume, indispensable to every medical man.

SAUNDERS' POCKET MEDICAL FORMULARY. By WILLIAM M. POWELL, M. D., Attending Physician to the Mercer House for Invalid Women at Atlantic City. Containing 1800 Formulæ, selected from several hundred of the best-known authorities. Forming a handsome and convenient pocket companion of nearly 300 printed pages, with blank leaves for Additions; with an Appendix containing Posological Table, Formulæ and Doses for Hypodermatic Medication, Poisons and their Antidotes, Diameters of the Female Pelvis and Fœtal Head, Obstetrical Table, Diet List for Various Diseases, Materials and Drugs used in Antiseptic Surgery, Treatment of Asphyxia from Drowning, Surgical Remembrancer, Tables of Incompatibles, Eruptive Fevers, Weights and Measures, etc. Handsomely bound in morocco, with side index, wallet, and flap. Price, $1.75 net.

FIFTH EDITION, THOROUGHLY REVISED.

"This little book, that can be conveniently carried in the pocket, contains an immense amount of material. It is very useful, and as the name of the author of each prescription is given, is unusually reliable."—*New York Medical Record.*

A COMPENDIUM OF INSANITY. By JOHN B. CHAPIN, M.D., LL.D., Physician-in-Chief, Pennsylvania Hospital for the Insane; late Physician-Superintendent of the Willard State Hospital, New York; Honorary Member of the Medico-Psychological Society of Great Britain, of the Society of Mental Medicine of Belgium. 12mo, 234 pages, illust. Cloth, $1.25 net.

The author has given, in a condensed and concise form, a compendium of Diseases of the Mind, for the convenient use and aid of physicians and students. It contains a clear, concise statement of the clinical aspects of the various abnormal mental conditions, with directions as to the most approved methods of managing and treating the insane.

"The practical parts of Dr. Chapin's book are what constitute its distinctive merit. We desire especially, however, to call attention to the fact that in the subject of the therapeutics of insanity the work is exceedingly valuable. The author has made a distinct addition to the literature of his specialty."—*Philadelphia Medical Journal.*

AN OPERATION BLANK, with Lists of Instruments, etc. required in Various Operations. Prepared by W. W. KEEN, M. D., LL.D., Professor of Principles of Surgery in the Jefferson Medical College, Philadelphia. Price per Pad, containing Blanks for fifty operations, 50 cents net.

SECOND EDITION, REVISED FORM.

A convenient blank, suitable for all operations, giving complete instructions regarding necessary preparation of patient, etc., with a full list of dressings and medicines to be employed.

On the back of each blank is a list of instruments used—viz. general instruments, etc., required for all operations; and special instruments for surgery of the brain and spine, mouth and throat, abdomen, rectum, male and female genito-urinary organs, the bones, etc.

The whole forming a neat pad, arranged for hanging on the wall of a surgeon's office or in the hospital operating-room.

"Will serve a useful purpose for the surgeon in reminding him of the details of preparation for the patient and the room as well as for the instruments, dressings, and antiseptics needed."—*New York Medical Record.*

"Covers about all that can be needed in any operation."—*American Lancet.*

"The plan is a capital one."—*Boston Medical and Surgical Journal.*

LABORATORY EXERCISES IN BOTANY. By EDSON S. BASTIN, M. A., Professor of Materia Medica and Botany in the Philadelphia College of Pharmacy. Octavo volume of 536 pages, 87 full-page plates. Price, Cloth, $2.50.

This work is intended for the beginner and the advanced student, and it fully covers the structure of flowering plants, roots, ordinary stems, rhizomes, tubers, bulbs, leaves, flowers, fruits, and seeds. Particular attention is given to the gross and microscopical structure of plants, and to those used in medicine. Illustrations have freely been used to elucidate the text, and a complete index to facilitate reference has been added.

"There is no work like it in the pharmaceutical or botanical literature of this country, and we predict for it a wide circulation."—*American Journal of Pharmacy.*

DIET IN SICKNESS AND IN HEALTH. By MRS. ERNEST HART, formerly Student of the Faculty of Medicine of Paris and of the London School of Medicine for Women; with an INTRODUCTION by Sir Henry Thompson, F. R. C. S., M. D., London. 220 pages; illustrated. Price, Cloth, $1.50.

Useful to those who have to nurse, feed, and prescribe for the sick. In each case the accepted causation of the disease and the reasons for the special diet prescribed are briefly described. Medical men will find the dietaries and recipes practically useful, and likely to save them trouble in directing the dietetic treatment of patients.

MANUAL OF PHYSIOLOGY, with Practical Exercises. For Students and Practitioners. By G. N. STEWART, M. A., M. D., D. Sc., lately Examiner in Physiology, University of Aberdeen, and of the New Museums, Cambridge University; Professor of Physiology in the Western Reserve University, Cleveland, Ohio. Handsome octavo volume of 848 pages, with 300 illustrations in the text, and 5 colored plates. Price, Cloth, $3.75 net.

THIRD EDITION, REVISED.

"It will make its way by sheer force of merit, and *amply deserves to do so. It is one of the very best English text-books on the subject.*"—*London Lancet.*

"Of the many text-books of physiology published, we do not know of one that so nearly comes up to the ideal as does Professor Stewart's volume."—*British Medical Journal.*

ESSENTIALS OF PHYSICAL DIAGNOSIS OF THE THORAX. By ARTHUR M. CORWIN, A. M., M. D., Demonstrator of Physical Diagnosis in the Rush Medical College, Chicago; Attending Physician to the Central Free Dispensary, Department of Rhinology, Laryngology, and Diseases of the Chest. 200 pages. Illustrated. Cloth, flexible covers. Price, $1.25 net.

SECOND EDITION, THOROUGHLY REVISED AND ENLARGED.

SYLLABUS OF OBSTETRICAL LECTURES in the Medical Department, University of Pennsylvania. By RICHARD C. NORRIS, A. M., M. D., Lecturer on Clinical and Operative Obstetrics, University of Pennsylvania. Third edition, thoroughly revised and enlarged. Crown 8vo. Price, Cloth, interleaved for notes, $2.00 net.

"This work is so far superior to others on the same subject that we take pleasure in calling attention briefly to its excellent features. It covers the subject thoroughly, and will prove invaluable both to the student and the practitioner. The author has introduced a number of valuable hints which would only occur to one who was himself an experienced teacher of obstetrics. The subject-matter is clear, forcible, and modern. We are especially pleased with the portion devoted to the practical duties of the accoucheur, care of the child, etc. The paragraphs on antiseptics are admirable; there is no doubtful tone in the directions given. No details are regarded as unimportant; no minor matters omitted. We venture to say that even the old practitioner will find useful hints in this direction which he cannot afford to despise."—*New York Medical Record.*

A SYLLABUS OF LECTURES ON THE PRACTICE OF SURGERY, arranged in conformity with "An American Text-Book of Surgery." By N. SENN, M. D., PH. D., Professor of Surgery in Rush Medical College, Chicago, and in the Chicago Polyclinic. Price, $2.00.

This work by so eminent an author, himself one of the contributors to "An American Text-Book of Surgery," will prove of exceptional value to the advanced student who has adopted that work as his text-book. It is not only the syllabus of an unrivalled course of surgical practice, but it is also an epitome of or supplement to the larger work.

"The author has evidently spared no pains in making his Syllabus thoroughly comprehensive, and has added new matter and alluded to the most recent authors and operations. Full references are also given to all requisite details of surgical anatomy and pathology."—*British Medical Journal*, London.

THE CARE OF THE BABY. By J. P. CROZER GRIFFITH, M. D., Clinical Professor of Diseases of Children, University of Pennsylvania; Physician to the Children's Hospital, Philadelphia, etc. 404 pages, with 67 illustrations in the text, and 5 plates. 12mo. Price, $1.50.

SECOND EDITION, REVISED.

A reliable guide not only for mothers, but also for medical students and practitioners whose opportunities for observing children have been limited.

"The whole book is characterized by rare good sense, and is evidently written by a master hand. It can be read with benefit not only by mothers, but by medical students and by any practitioners who have not had large opportunities for observing children."—*American Journal of Obstetrics.*

THE NURSE'S DICTIONARY of Medical Terms and Nursing Treatment, containing Definitions of the Principal Medical and Nursing Terms, Abbreviations, and Physiological Names, and Descriptions of the Instruments, Drugs, Diseases, Accidents, Treatments, Operations, Foods, Appliances, etc. encountered in the ward or the sick-room. By HONNOR MORTEN, author of "How to Become a Nurse," "Sketches of Hospital Life," etc. 16mo, 140 pages. Price, Cloth, $1.00.

This little volume is intended for use merely as a small reference-book which can be consulted at the bedside or in the ward. It gives sufficient explanation to the nurse to enable her to comprehend a case until she has leisure to look up larger and fuller works on the subject.

DIET LISTS AND SICK-ROOM DIETARY. By JEROME B. THOMAS, M. D., Visiting Physician to the Home for Friendless Women and Children and to the Newsboys' Home; Assistant Visiting Physician to the Kings County Hospital; Assistant Bacteriologist, Brooklyn Health Department. Price, Cloth, $1.50 (Send for specimen List.)

One hundred and sixty detachable (perforated) diet lists for Albuminuria, Anæmia and Debility, Constipation, Diabetes, Diarrhœa, Dyspepsia, Fevers, Gout or Uric-Acid Diathesis, Obesity, and Tuberculosis. Also forty detachable sheets of Sick-Room Dietary, containing full instructions for preparation of easily-digested foods necessary for invalids. Each list is *numbered only*, the disease for which it is to be used in no case being mentioned, an index key being reserved for the physician's private use.

DIETS FOR INFANTS AND CHILDREN IN HEALTH AND IN DISEASE. By LOUIS STARR, M. D., Editor of "An American Text-Book of the Diseases of Children." 230 blanks (pocket-book size) perforated and neatly bound in flexible morocco. Price, $1.25 net.

The first series of blanks are prepared for the first seven months of infant life; each blank indicates the ingredients, but not the *quantities*, of the food, the latter directions being left for the physician. After the seventh month, modifications being less necessary, the diet lists are printed in full. Formula for the preparation of diluents and foods are appended.

HOW TO EXAMINE FOR LIFE INSURANCE. By JOHN M. KEATING, M. D., Fellow of the College of Physicians and Surgeons of Philadelphia; Vice-President of the American Pædiatric Society; Ex-President of the Association of Life Insurance Medical Directors. Royal 8vo, 211 pages, with two large half-tone illustrations, and a plate prepared by Dr. McClellan from special dissections; also, numerous cuts to elucidate the text. Third edition. Price, Cloth, $2.00 net.

"This is by far the most useful book which has yet appeared on insurance examination, a subject of growing interest and importance. Not the least valuable portion of the volume is Part II., which consists of instructions issued to their examining physicians by twenty-four representative companies of this country. As the proofs of these instructions were corrected by the directors of the companies, they form the latest instructions obtainable. If for these alone, the book should be at the right hand of every physician interested in this special branch of medical science."—*The Medical News*, Philadelphia.

NURSING: ITS PRINCIPLES AND PRACTICE. By ISABEL ADAMS HAMPTON, Graduate of the New York Training School for Nurses attached to Bellevue Hospital; Superintendent of Nurses and Principal of the Training School for Nurses, Johns Hopkins Hospital, Baltimore, Md.; late Superintendent of Nurses, Illinois Training School for Nurses, Chicago, Ill. In one very handsome 12mo volume of 512 pages, illustrated. Price, Cloth, $2.00 net.

SECOND EDITION, REVISED AND ENLARGED.

This original work on the important subject of nursing is at once comprehensive and systematic. It is written in a clear, accurate, and readable style, suitable alike to the student and the lay reader. Such a work has long been a desideratum with those entrusted with the management of hospitals and the instruction of nurses in training-schools. It is also of especial value to the graduated nurse who desires to acquire a practical working knowledge of the care of the sick and the hygiene of the sick-room.

OBSTETRIC ACCIDENTS, EMERGENCIES, AND OPERATIONS. By L. CH. BOISLINIERE, M. D., late Emeritus Professor of Obstetrics in the St. Louis Medical College. 381 pages, handsomely illustrated. Price, $2.00 net.

"For the use of the practitioner who, when away from home, has not the opportunity of consulting a library or of calling a friend in consultation. He then, being thrown upon his own resources, will find this book of benefit in guiding and assisting him in emergencies."

INFANT'S WEIGHT CHART. Designed by J. P. CROZER GRIFFITH, M. D., Clinical Professor of Diseases of Children in the University of Pennsylvania. 25 charts in each pad. Price per pad, 50 cents net.

A convenient blank for keeping a record of the child's weight during the first two years of life. Printed on each chart is a curve representing the average weight of a healthy infant, so that any deviation from the normal can readily be detected

SAUNDERS' NEW SERIES OF MANUALS

for Students and Practitioners.

THAT there exists a need for thoroughly reliable hand-books on the leading branches of Medicine and Surgery is a fact amply demonstrated by the favor with which the SAUNDERS NEW SERIES OF MANUALS have been received by medical students and practitioners and by the Medical Press. These manuals are not merely condensations from present literature, but are ably written by well-known authors and practitioners, most of them being teachers in representative American colleges. Each volume is concisely and authoritatively written and exhaustive in detail, without being encumbered with the introduction of "cases," which so largely expand the ordinary text-book. These manuals will therefore form an admirable collection of advanced lectures, useful alike to the medical student and the practitioner: to the latter, too busy to search through page after page of elaborate treatises for what he wants to know, they will prove of inestimable value; to the former they will afford safe guides to the essential points of study.

The SAUNDERS NEW SERIES OF MANUALS are conceded to be superior to any similar books now on the market. No other manuals afford so much information in such a concise and available form. A liberal expenditure has enabled the publisher to render the mechanical portion of the work worthy of the high literary standard attained by these books.

Any of these Manuals will be mailed on receipt of price (see next page for List).

SAUNDERS' NEW SERIES OF MANUALS.

VOLUMES PUBLISHED.

PHYSIOLOGY. By JOSEPH HOWARD RAYMOND, A. M., M. D., Professor of Physiology and Hygiene and Lecturer on Gynecology in the Long Island College Hospital, etc. Price, $1.25 net.

SURGERY, General and Operative. By JOHN CHALMERS DACOSTA, M. D., Professor of Clinical Surgery, Jefferson Medical College, Philadelphia. Second edition, revised and greatly enlarged. Octavo, 911 pages, 386 illustrations. Cloth, $4.00 net; Half-Morocco, $5.00 net.

DOSE-BOOK AND MANUAL OF PRESCRIPTION-WRITING. By E. Q. THORNTON, M. D., Demonstrator of Therapeutics, Jefferson Medical College, Philadelphia. Price, $1.25 net.

MEDICAL JURISPRUDENCE. By HENRY C. CHAPMAN, M. D., Professor of Institutes of Medicine and Medical Jurisprudence in the Jefferson Medical College of Philadelphia, etc. Price, $1.50 net.

SURGICAL ASEPSIS. By CARL BECK, M. D., Surgeon to St. Mark's Hospital and to the German Poliklinik; Instructor in Surgery, New York Post-Graduate Medical School, etc. Price, $1.25 net.

MANUAL OF ANATOMY. By IRVING S. HAYNES, M. D., Adjunct Professor of Anatomy and Demonstrator of Anatomy, Medical Department of the New York University, etc. Price, $2.50 net.

SYPHILIS AND THE VENEREAL DISEASES. By JAMES NEVINS HYDE, M. D., Professor of Skin and Venereal Diseases, and FRANK H. MONTGOMERY, M. D., Lecturer on Dermatology and Genitourinary Diseases in Rush Medical College, Chicago. Price, $2.50 net.

PRACTICE OF MEDICINE. By GEORGE ROE LOCKWOOD, M. D., Professor of Practice in the Woman's Medical College of the New York Infirmary, etc. Price, $2.50 net.

OBSTETRICS. By W. A. NEWMAN DORLAND, M. D., Assistant Demonstrator of Obstetrics, University of Pennsylvania; Chief of Gynecological Dispensary, Pennsylvania Hospital. Price, $2.50 net.

DISEASES OF WOMEN. By J. BLAND SUTTON, F. R. C. S., Assistant Surgeon to the Middlesex Hospital, and Surgeon to the Chelsea Hospital for Women, London; and ARTHUR E. GILES, M. D., B. Sc. Lond., F. R. C. S. Edin., Assistant Surgeon to the Chelsea Hospital for Women, London. 436 pages, handsomely illustrated. Price, $2.50 net.

IN PREPARATION.

NERVOUS DISEASES. By CHARLES W. BURR, M. D., Clinical Professor of Nervous Diseases, Medico-Chirurgical College, Philadelphia, etc.

∗ There will be published in the same series, at short intervals, carefully prepared works on various subjects, by prominent specialists.

SAUNDERS' QUESTION COMPENDS.

Arranged in Question and Answer Form.

THE LATEST, MOST COMPLETE, and BEST ILLUSTRATED SERIES OF COMPENDS EVER ISSUED.

Now the Standard Authorities in Medical Literature

WITH

Students and Practitioners in every City of the United States and Canada.

THE REASON WHY.

They are the advance guard of "Student's Helps"—that DO HELP; they are the leaders in their special line, *well and authoritatively written by able men, who, as teachers in the large colleges, know exactly what is wanted by a student preparing for his examinations.* The judgment exercised in the selection of authors is fully demonstrated by their professional elevation. Chosen from the ranks of Demonstrators, Quiz-masters, and Assistants, most of them have become Professors and Lecturers in their respective colleges.

Each book is of convenient size (5 × 7 inches), containing on an average 250 pages, profusely illustrated, and elegantly printed in clear, readable type, on fine paper.

The entire series, numbering twenty-four subjects, has been kept thoroughly revised and enlarged when necessary, many of them being in their fourth and fifth editions.

TO SUM UP.

Although there are numerous other Quizzes, Manuals, Aids, etc. in the market, none of them approach the "Blue Series of Question Compends;" and the claim is made for the following points of excellence:

1. Professional distinction and reputation of authors.
2. Conciseness, clearness, and soundness of treatment.
3. Size of type and quality of paper and binding.

* Any of these Compends will be mailed on receipt of price (see next page for List).

SAUNDERS' QUESTION-COMPEND SERIES.

Price, Cloth, $1.00 per copy, except when otherwise noted.

1. **ESSENTIALS OF PHYSIOLOGY.** 4th edition. Illustrated. Revised and enlarged. By H. A. Hare, M. D. (Price, $1.00 net.)
2. **ESSENTIALS OF SURGERY.** 6th edition, with an Appendix on Antiseptic Surgery. 90 illustrations. By Edward Martin, M. D.
3. **ESSENTIALS OF ANATOMY.** 6th edition, thoroughly revised. 151 illustrations. By Charles B. Nancrede, M. D.
4. **ESSENTIALS OF MEDICAL CHEMISTRY, ORGANIC AND INORGANIC.** 5th edition, revised, with an Appendix. By Lawrence Wolff, M. D.
5. **ESSENTIALS OF OBSTETRICS.** 4th edition, revised and enlarged. 75 illustrations. By W. Easterly Ashton, M. D.
6. **ESSENTIALS OF PATHOLOGY AND MORBID ANATOMY.** 7th thousand. 46 illustrations. By C. E. Armand Semple, M. D.
7. **ESSENTIALS OF MATERIA MEDICA, THERAPEUTICS, AND PRESCRIPTION-WRITING.** 5th edition. By Henry Morris, M. D.
8, 9. **ESSENTIALS OF PRACTICE OF MEDICINE.** By Henry Morris, M. D. An Appendix on Urine Examination. Illustrated. By Lawrence Wolff, M. D. 3d edition, enlarged by some 300 Essential Formulæ, selected from eminent authorities, by Wm. M. Powell, M. D. (Double number, price $2.00.)
10. **ESSENTIALS OF GYNÆCOLOGY.** 4th edition, revised. With 62 illustrations. By Edwin B. Cragin, M. D.
11. **ESSENTIALS OF DISEASES OF THE SKIN.** 4th edition, revised and enlarged. 71 letter-press cuts and 15 half-tone illustrations. By Henry W. Stelwagon, M. D. (Price, $1.00 net.)
12. **ESSENTIALS OF MINOR SURGERY, BANDAGING, AND VENEREAL DISEASES.** 2d edition, revised and enlarged. 78 illustrations. By Edward Martin, M. D.
13. **ESSENTIALS OF LEGAL MEDICINE, TOXICOLOGY, AND HYGIENE.** 130 illustrations. By C. E. Armand Semple, M. D.
14. **ESSENTIALS OF DISEASES OF THE EYE, NOSE, AND THROAT.** 124 illustrations. 2d edition, revised. By Edward Jackson, M. D., and E. Baldwin Gleason, M. D.
15. **ESSENTIALS OF DISEASES OF CHILDREN.** 2d edition. By William M. Powell, M. D.
16. **ESSENTIALS OF EXAMINATION OF URINE.** Colored "Vogel Scale," and numerous illustrations. By Lawrence Wolff, M. D. (Price, 75 cents.)
17. **ESSENTIALS OF DIAGNOSIS.** 55 illustrations, some in colors. By S. Solis-Cohen, M. D., and A. A. Eshner, M. D. (Price, $1.50 net.)
18. **ESSENTIALS OF PRACTICE OF PHARMACY.** 2d edition, revised. By L. E. Sayre.
20. **ESSENTIALS OF BACTERIOLOGY.** 3d edition. 82 illustrations. By M. V. Ball, M. D.
21. **ESSENTIALS OF NERVOUS DISEASES AND INSANITY.** 48 illustrations. 3d edition, revised. By John C. Shaw, M. D.
22. **ESSENTIALS OF MEDICAL PHYSICS.** 155 illustrations. 2d edition, revised. By Fred J. Brockway, M. D. (Price, $1.00 net.)
23. **ESSENTIALS OF MEDICAL ELECTRICITY.** 65 illustrations. By David D. Stewart, M. D., and Edward S. Lawrance, M. D.
24. **ESSENTIALS OF DISEASES OF THE EAR.** 114 illustrations. 2d edition, revised and enlarged. By E. Baldwin Gleason, M. D.

IN PRESS

FOR PUBLICATION EARLY IN THE FALL OF 1899.

THE INTERNATIONAL TEXT-BOOK OF SURGERY. In two vols. By American and British authors. Edited by J. COLLINS WARREN, M. D., LL.D., Professor of Surgery, Harvard Medical School, Boston; Surgeon to the Massachusetts General Hospital; and A. PEARCE GOULD, M. S., F. R. C. S., Eng., Lecturer on Practical Surgery and Teacher of Operative Surgery, Middlesex Hospital Medical School; Surgeon to the Middlesex Hospital, London, England. Vol. I. Handsome octavo volume of about 950 pages, with over 400 beautiful illustrations in the text, and 9 lithographic plates.

HEISLER'S EMBRYOLOGY.

A Text-Book of Embryology. By JOHN C. HEISLER, M. D., Professor of Anatomy in the Medico-Chirurgical College, Philadelphia. 12mo volume of about 325 pages, handsomely illustrated.

KYLE ON THE NOSE AND THROAT.

Diseases of the Nose and Throat. By D. BRADEN KYLE, M. D., Clinical Professor of Laryngology and Rhinology, Jefferson Medical College, Philadelphia; Consulting Laryngologist, Rhinologist, and Otologist, St. Agnes' Hospital. Octavo volume of about 630 pages, with over 150 illustrations and 6 lithographic plates.

PRYOR PELVIC INFLAMMATIONS.

The Treatment of Pelvic Inflammations through the Vagina. By W. R. PRYOR, M. D., Professor of Gynecology in the New York Polyclinic. 12mo volume of about 250 pages, handsomely illustrated.

ABBOTT ON TRANSMISSIBLE DISEASES.

The Hygiene of Transmissible Diseases: their Causation, Modes of Dissemination, and Methods of Prevention. By A. C. ABBOTT, M. D., Professor of Hygiene in the University of Pennsylvania; Director of the Laboratory of Hygiene. Octavo volume of about 325 pages, containing a number of charts and maps, and numerous illustrations.

JACKSON—DISEASES OF THE EYE.

A Manual of Diseases of the Eye. By EDWARD JACKSON, A. M., M. D., late Professor of Diseases of the Eye in the Philadelphia Polyclinic and College for Graduates in Medicine. 12mo volume of over 500 pages, with about 175 beautiful illustrations from drawings by the author.

www.ingramcontent.com/pod-product-compliance
Lightning Source LLC
Chambersburg PA
CBHW030010240426
43672CB00007B/894